Technikzukünfte, Wissenschaft und Gesellschaft / Futures of Technology, Science and Society

Herausgegeben von
A. Grunwald, Karlsruhe, Germany
R. Heil, Karlsruhe, Germany
C. Coenen, Karlsruhe, Germany

Diese interdisziplinäre Buchreihe ist Technikzukünften in ihren wissenschaftlichen und gesellschaftlichen Kontexten gewidmet. Der Plural „Zukünfte" ist dabei Programm. Denn erstens wird ein breites Spektrum wissenschaftlich-technischer Entwicklungen beleuchtet, und zweitens sind Debatten zu Technowissenschaften wie u. a. den Bio-, Informations-, Nano-und Neurotechnologien oder der Robotik durch eine Vielzahl von Perspektiven und Interessen bestimmt. Diese Zukünfte beeinflussen einerseits den Verlauf des Fortschritts, seine Ergebnisse und Folgen, z.B. durch Ausgestaltung der wissenschaftlichen Agenda. Andererseits sind wissenschaftlich-technische Neuerungen Anlass, neue Zukünfte mit anderen gesellschaftlichen Implikationen auszudenken. Diese Wechselseitigkeit reflektierend, befasst sich die Reihe vorrangig mit der sozialen und kulturellen Prägung von Naturwissenschaft und Technik, der verantwortlichen Gestaltung ihrer Ergebnisse in der Gesellschaft sowie mit den Auswirkungen auf unsere Bilder vom Menschen.

This interdisciplinary series of books is devoted to technology futures in their scientific and societal contexts. The use of the plural "futures" is by no means accidental: firstly, light is to be shed on a broad spectrum of developments in science and technology; secondly, debates on technoscientific fields such as biotechnology, information technology, nanotechnology, neurotechnology and robotics are influenced by a multitude of viewpoints and interests. On the one hand, these futures have an impact on the way advances are made, as well as on their results and consequences, for example by shaping the scientific agenda. On the other hand, scientific and technological innovations off er an opportunity to conceive of new futures with different implications for society. Reflecting this reciprocity, the series concentrates primarily on the way in which science and technology are influenced social and culturally, on how their results can be shaped in a responsible manner in society, and on the way they affect our images of humankind.

Prof. Dr. Armin Grunwald, Physiker, Mathematiker und Philosoph, lehrt Technikphilosophie und Technikethik am Karlsruher Institut für Technologie (KIT), ist Leiter des Instituts für Technikfolgenabschätzung und Systemanalyse (ITAS) in Karlsruhe und Leiter des Büros für Technikfolgen-Abschätzung beim Deutschen Bundestag (TAB) in Berlin. / Professor Armin Grunwald, physicist, mathematician and philosopher, teaches the philosophy and ethics of technology at the Karlsruhe Institute of Technology (KIT), and is the director of the Institute for Technology Assessment and Systems Analysis (ITAS) in Karlsruhe and of the Office of Technology Assessment at the German Bundestag (TAB) in Berlin.

Reinhard Heil, Philosoph, ist wissenschaftlicher Mitarbeiter am KIT-ITAS. / Reinhard Heil, philosopher, is a researcher at KIT-ITAS.

Christopher Coenen, Politikwissenschaftler, ist wissenschaftlicher Mitarbeiter am KIT-ITAS und Herausgeberder Zeitschrift ‚NanoEthics: Studies of New and Emerging Technologies'. / Christopher Coenen, political scientist, is a researcher at KIT-ITAS and the editor-in-chief of the journal ‚NanoEthics: Studies of New and Emerging Technologies'.

More information about this series at http://www.springer.com/series/13596

Reinhard Heil · Stefanie B. Seitz
Harald König · Jürgen Robienski
Editors

Epigenetics

Ethical, Legal and Social Aspects

 Springer VS

Editors
Reinhard Heil
Inst. Technikfolgenab. u. Systemanalyse
Karlsruher Institut für Technologie
Karlsruhe
Germany

Stefanie B. Seitz
DBFZ – Deutsches
Biomasseforschungszentrum gGmbH
Leipzig
Germany

Harald König
Inst. Technikfolgenab. u. Systemanalyse
Karlsruher Institut für Technologie
Karlsruhe
Germany

Jürgen Robienski
Müden
Germany

Technikzukünfte, Wissenschaft und Gesellschaft / Futures of Technology,
Science and Society
ISBN 978-3-658-14459-3   ISBN 978-3-658-14460-9   (eBook)
DOI 10.1007/978-3-658-14460-9

Library of Congress Control Number: 2016946939

Lektorat: Frank Schindler

Printed on acid-free paper

This Springer VS imprint is published by Springer Nature
The registered company is Springer Fachmedien Wiesbaden GmbH
The registered company address is: Abraham-Lincoln-Strasse 46, 65189 Wiesbaden, Germany

Contents

Introduction

Reinhard Heil, Stefanie B. Seitz, Jürgen Robienski and Harald König

Abstract

Within the last decade, epigenetics has developed as a branch of molecular genetics and is currently experiencing a true hype. One of the most fascinating but also scaring findings of epigenetics is the confirmation of the old presumption that the environment can have lasting effects on gene expression – maybe even into the next generation(s). Thus, epigenetics has not only hugely expanded the horizon concerning the understanding of regulation mechanisms that influence the appearance of organisms, epigenetic knowledge pools also have a huge potential for innovations – especially in the health care sector (e.g. fighting cancer, Alzheimer's and diabetes). But epigenetic research also provides advice for "better living" in respect to diet, everyday behavior, social relations and many more. This volume illuminates ethical, legal and social

R. Heil (✉) · H. König
Institute for Technology Assessment and Systems Analysis (ITAS), KIT—The Research University in the Helmholtz Association, Karlstraße 11, 76133 Karlsruhe, Germany
e-mail: Reinhard.Heil@kit.edu

H. König
e-mail: h.koenig@kit.edu

S.B. Seitz
DBFZ – Deutsches Biomasseforschungszentrum gGmbH,
Torgauer Str. 116, 04347 Leipzig, Germany
e-mail: stefanie.seitz@dbfz.de

J. Robienski
Eichenkamp 6, Müden, Germany
e-mail: robienski@aol.com

© Springer Fachmedien Wiesbaden GmbH 2017
R. Heil et al. (eds.), *Epigenetics*, Technikzukünfte, Wissenschaft und Gesellschaft / Futures of Technology, Science and Society,
DOI 10.1007/978-3-658-14460-9_1

1

aspects of epigenetics from multidisciplinary perspectives that have been given little consideration to date. Here we will introduce its contributions and put them into context.

Within the last decade, epigenetics has developed as a branch of molecular genetics and is currently experiencing a true hype. Modern epigenetics unites scientists from life sciences, organic chemistry as well as computer and engineering sciences to find an answer to the question of how environmental influences can have a lasting effect on gene expression, maybe even into the next generations. In doing so, the focus is mainly on the following three processes: retroactive modification of specific DNA bases (e.g. DNA methylations), changes of the chromatin (e.g. through modifications of histone composition and structure), and RNA-mediated gene regulation mechanisms (e.g. so-called RNA interference) (for an overview see e.g. in Youngson and Whitelaw 2008).

Epigenetics has not only hugely expanded the horizon concerning the understanding of regulation mechanisms that influence the appearance of organisms, it has also developed a new, more comprehensive picture of heritability and finally also of evolution theory. The fundamental ascertainment that a changed phenotype is also observed in subsequent generations, and was therefore inherited without a detectable corresponding change in the gene sequence, makes it tempting to call epigenetics the "Science of Change" (e.g. Weinhold 2006). There is still lively discussion upon several of Jean Baptiste Lamarck's (1744–1829) outdated theories that were dragged out of the scientific history's dustbin. Still, it is indisputable that even if there are currently no final definitions of which (molecular) biological phenomena should be included under "epigenetics" and which not, epigenetics is perceived as revolutionary in the field of life sciences. Epigenetics shares this definition imprecision with other young disciplines of science and emerging technologies (e.g. nanotechnology or synthetic biology, cf. Fleischer and Quendt 2007; König et al. 2013). At the present time, the term "epigenetics" still often serves as a sort of "key umbrella term"[1] and therefore helps the parties involved in research and development activities in this field to act strategically in terms of research funding politics.

In addition to their significance for basic research, epigenetic knowledge pools also have a huge potential for innovations—and many groups of players are hoping

---

[1]In the literature, these are also called "boundary objects" (Star and Griesemer 1989; Guston 2001) or "umbrella terms" (Rip and Voß 2013). These are (generic) terms, which, on the one hand, are open enough for use by different groups and allow for their own interpretations, but on the other, are defined narrowly enough that they still preserve a "global identity" and ensure a shared understanding.

for progress in their fields. Especially for health professionals, who want to make headway in the battle against big endemic diseases of the western world (including cancer, Alzheimer's and diabetes) with the help of epigenetics. This naturally also raises the hopes of patients and their relatives. Furthermore, epigenetic research has provided strong indications that our everyday behaviour (e.g. diet or substance consumption), social experiences (e.g. motherly care, traumatic experiences, or stress), certain chemicals and other environmental influences can have effects on the epigenome. In other words, they can be reflected in the epigenetic markings on our genome or can leave their traces there (cf. Weinhold 2006). In addition, several of these epigenetic markings seem to be heritable and therefore even have an influence on the next generations (cf. Grossniklaus et al. 2013). A more recent theory also relates the development of homosexuality with epigenetic mechanisms (cf. Rice et al. 2012). But also beyond the therapy and diagnostics of diseases such as cancer or diabetes, knowledge of epigenetics already has a growing influence on our lives, for example, with the question regarding the influence of environmental conditions on our behaviour and the consequences of in vitro fertilisation (van Montfoort et al. 2012) or the reproductive clones used in animal breeding (Smith et al. 2012).

At the same time, the consequences of these findings for our society and each of our daily lives still remain largely unexplored. Until now, only few scientific investigations have been published on the legal, ethical and social effects of epigenetics (e.g. Rothstein et al. 2009; Rothstein 2013) and the interest in the corresponding disciplines outside of life sciences is rather limited to date. There are still many open questions for research regarding the definition and scope of the concept of epigenetics as well as the identification of relevant fields and of specific problems relating to epigenetics. The BMBF summer school, the results of which are presented here, deals specifically with the identification of these legal, ethical and social implications.[2] Until now, the following questions have been discussed, which were then also examined by the authors of this publication and analysed in detail:

[2]The articles in this publication are the result of the conference week "Epigenetics—Ethical, Legal and Social Aspects", generously funded by the German Federal Ministry of Education and Research (grant nr. 01GP1282), which was implemented in 2013 by the editors at the Institute for Technology Assessment and Systems Analysis (ITAS) of the Karlsruhe Institute of Technology (KIT). We would like to warmly thank Claudia Brändle, Muazez Genc and Martin Sand for their support with the preparation and implementation of the conference week.

- Regulation of research as well of the question of responsible handling of epigenetic knowledge pools and existing "ignorance",[3]
- regulation of health, environmental and work protection,
- legal consequences in terms of personal injury and medical malpractice, but also personal responsibility, privacy and data security,
- justice questions such as equality and discrimination, access to healthcare as well as inter-generation fairness, and
- other normative questions, e.g. regarding (personal) responsibility.

Altogether, the articles in this publication aim to contribute to answering the question regarding the significance that knowledge pools from epigenetic research have or could have for society, the possible interpretations, and whether they are suitable for leading to new mindsets and changes in values.

This publication starts with the article *"Introduction to Epigenetics"* by Jörn Walter and Anja Hümpel. The epigeneticist and his co-author provide an introduction to the basic principles of epigenetics and present the current state of research in understandable terms and using concrete examples.

In his article *"Too Early or Too Late? The Assessment of Emerging Technology"*, Michael Decker addresses the question of which topics are being looked at by technology assessment research and when. Technology assessment currently faces the dilemma that there is an area of conflict between the early and later studies of technology impacts. If one wants to shape technology, one has to think about the possible consequences very early on, but at this point in time, very little knowledge is available. If one waits until sufficient knowledge is available, the technical development process has already progressed so far that there are hardly any options left for shaping its outcome. Technology assessment (TA) has responded methodologically to the different challenges and in particular has developed a tool set for assessing what is called the new and emerging technologies (NEST). Decker presents these criteria and uses them to derive consequences for the study of epigenetics by the TA.

---

[3]Here, "ignorance" does not mean "not knowing yet", which "could be transformed into knowledge with further research", but rather "irreducible ignorance, which inevitably arises with questions that are continuously reformulated by science regarding the future of society and the associated long-term planning horizons. This is typical e.g. for the assessment of risks and consequences of new scientific-technical methods, which is decisive for political action, for which science provides at the most plausible assumptions, but no validated knowledge" (Hennen et al. 2004, p. 19; own translation).

Epigenetic research results are often cited as a rebuttal of so-called gene determinism. In his article *"Epigenetics and Genetic Determinism (in Popular Science)"*, Sebastian Schuol investigates the relation of epigenetics and gene determinism as antagonistic concepts. Whether epigenetics refutes gene determinism depends on the basic concept of "information". Although the argument of gene-environment interaction refutes a simple version of gene determinism, a more complex version still remains. According to Schuol's theory, the supposed solution even subliminally promotes the gene deterministic concept. Complex gene determinism arises when genetic information is considered as an instruction in an intentional sense, because the epigenetic information then only serves for its regulation. For this reason, the reference to the gene/environment interaction is not sufficient for a full rebuttal of gene determinism.

Specific ethical problems that are associated with epigenetics are addressed in the article *"Identity and Non-Identity. Intergenerational Justice as a Topic of an Ethics of Epigenetics"* by Phillip Bode. Following the so-called Överkalix studies (Kaati et al. 2002), which demonstrated that, among other things, malnutrition occurring in a specific period in boys' lives apparently has a weakening effect on the risk of cardiovascular disease for their direct progeny, Bode investigates whether one should subject one's own child to malnutrition, even only slightly, in order to reduce the (statistical) risk of disease of possible great-grandsons, or if one should even be ethically obligated to do so.

In his article *"Genetics, Epigenetics and Forms of Actions. About the Ethical Ambivalence of Epigenetic Knowledge"*, Joachim Boldt demonstrates that epigenetics calls for an expansion of the discussion about the ethical aspects of genetics in two regards. On the one hand, the inclusion of environmental factors causes the relationship of the individual that is provided with predictive diagnoses to change with this knowledge. On the other hand, this correlation leads to the question as to how the instrumentally useful epigenetic findings for the patient can be embedded in the applicable context of social-communicative action that is relevant for the patient's everyday life.

Kirsten Brukamp's article *"Epigenetics: Biological, Medical, Social, and Ethical Challenges"* identifies a series of difficulties that are associated with the constantly increasing epigenetic knowledge pools in biology and their implications for sociology and ethics. In terms of the disciplines of biology and medicine, it is difficult to define the precise theoretical status of epigenetics in the nexus between nucleotide sequence and gene regulation. From a socio-scientific view, there are large uncertainties with regards to ignorance and inability as sociological categories. The topics of knowledge and risk communication, intergenerational justice, reproduction as well as social and public responsibility are of great ethical significance. The question arises as to how—taking account of the uncertain knowledge pools of epigenetics—the population must and can be informed.

In her article *"Learning from and Shaping the Public Discourse about Epigenetics"*, Stefanie B. Seitz demonstrates the significance of investigating public perception for the assessment of the consequences of new sciences and technologies, and which lessons can (and should) be learned from this. Because not only the scientific, but also the public discourse is instigated and is present in the media, her article questions the extent to which scientific actors should deliberately shape this discourse through active participation. This assumes that the discourse is inevitably influenced by epigeneticists through the publication of their results as well as by researchers from the field of humanities and social sciences through their activities. The goal of deliberate participation in the public discourse could be to contribute to finding a socially responsible handling of this field of knowledge and its applications, along the lines of Responsible Research and Innovation (RRI).[4]

In the article by Stefanie B. Seitz and Sebastian Schuol empirical base data on the *"State of the Public Discourse on Epigenetics"* is introduced for the first time in German-speaking countries. The article presents three case studies: First, a media analysis investigating the communicator side; second, a discourse analysis recording the position of the recipient, and third, the assessment of a public event on epigenetics.[5] It was observed that especially the aspect of individual responsibility plays a role in the public discourse, and that this discourse has been addressed very cautiously until now, and mainly in the media.

Jutta Jahnel's article *"Epigenetics—New Aspects of Chemicals Policy"* poses the question as to whether the existing regulations for chemicals take enough consideration of the known and possible epigenetic action mechanisms. A central point of her deliberations are the hormone-disturbing effects of so-called endocrine disruptors, like for example Bisphenol A. Jahnel sees the consequences of human-related changes in the environment, which have an effect on epigenetic mechanisms, as a challenge that can only be conquered through close collaboration of science, politics and ethics.

---

[4]Since several years, RRI is being discussed as a conceptual framework for the shaping and funding of innovation processes—with the goal of better adapting innovations to societal needs through the continuous involvement of various groups of actors in the innovation process right from the beginning, and therefore making them more successful. The RRI concept was mainly introduced by the European Commission (von Schomberg 2013; EC 2013) and was seized and further developed by TA researchers (Owen et al. 2013).

[5]This was the closing event of the BMBF conference week, the results of which are presented in this publication.

The article *"Epigenetics and Legal regulations: A Challenge for the State's Obligation to Protect Environment, Individual's Health and Civil Rights"* by Jürgen Robienski explores which fields of law could be significant for epigenetics. What are the consequences of epigenetic findings e.g. for the risk assessment of new and old technologies and noxa? Who is responsible and who is liable in the case of epigenetically relevant environmental influences? Are epigenetic changes to be considered as health impairment, which can justify civil claims for compensation or claims for benefits based on social law? What is the significance of epigenetically relevant environmental influences for public health protection in the form of precaution and prevention, and for work, maternity and child protection?

Also from a legal perspective, in her article *"Epigenetics and the Protection of Personality Rights"*, Caroline Fündling explores whether data about epigenetic states have the same personality relevance as genetic data and whether they call for a new evaluation of existing protection concepts. Because it is principally invariable, genetic data can have an influence on reproductive decisions and can be relevant not only for the directly affected person, but also for his or her relatives. For this reason, it is considered to be particularly worthy of protection. On the one hand, genetic data is covered by the right to informational self-determination, and on the other, the so-called right to ignorance should protect from the burdening contents of genetic information. This article explores whether this should also be applicable for epigenetic data.

Adiposity (obesity) is known to be one of the most momentous lifestyle diseases. For this reason, adiposity research is particularly interested in epigenetic approaches, because the activity of genes that are linked with the development of adiposity could potentially be influenced epigenetically, e.g. through dietary habits. In his article *"Adam's Apple and His Legacy: Ethical Perspectives on Epigenetics with an Excursion to the Field of Body Weight Regulation"*, Jens Ried pursues social-ethical questions regarding justice and equality in healthcare regarding the relationship between the genetic diagnostic findings and personal responsibility.

Harald Matern's article *"Epigenetics and Original Sin. Theological-Ethical Reflections on Heredity and Responsibility"* joins two theological-ethical perspectives on epigenetics to create a responsible-ethical scheme. He looks at which aspects of the topic are related to Christian tradition. Hereby, he is focussing on the one hand on the public-medial reception, and on the other on the structural proximity of several aspects of epigenetics to Christian teachings about original sin. The latter represents a central component of theological anthropology. Responsibility plays a central role in Christian teachings of original sin. The concept of responsibility can be formulated more precisely through the discussion about epigenetics. This makes it possible to promote a public discourse that is no longer based on moral perspectives, a key aspect especially for a theological-ethical position.

References

EC (European Commission). (2013). Options for strengthening responsible research and innovation. Brussels: Directorate-General for Research and Innovation Science in Society, European Commission (EUR25766). http://ec.europa.eu/research/science-society/document_library/pdf_06/options-for-strengthening_en.pdf

Fleischer, T., & Quendt, C. (2007). *Unsichtbar und unendlich – Bürgerperspektiven auf Nano-partikel - Ergebnisse zweier Fokusgruppen-Veranstaltungen in Karlsruhe.* Karlsruhe: Wissenschaftliche Berichte (FZKA 7337).

Grossniklaus, U., Kelly, B., Ferguson-Smith, A. C., Pembrey, M., & Lindquist, S. (2013). Transgenerational epigenetic inheritance: How important is it? *Nature Review Genetics, 14*(3), 228–235.

Guston, D. H. (2001). Boundary organizations in environmental policy and science: An introduction. *Science, Technology and Human Values, 26*(4), 399–408.

Hennen, L., Petermann, Th., & Scherz, C. (Eds.) (2004). *Partizipative Verfahren der Technikfolgen-Abschätzung und parlamentarische Politikberatung. Neue Formen der Kommunikation zwischen Wissenschaft, Politik und Öffentlichkeit.* Berlin: Büro für Technikfolgen-Abschätzung beim Deutschen Bundestag (TAB), TAB-Arbeitsbericht Nr. 96.

Kaati, G., Bygren, L. O., & Edvinsson, S. (2002). Cardiovascular and diabetes mortality determined by nutrition during parents' and grandparents' slow growth period. *European Journal of Human Genetics, 10*, 682–688.

König, H., Frank, D., Heil, R., & Coenen, C. (2013). Synthetic genomics and synthetic biology applications between hopes and concerns. *Current Genomics, 14*(1), 11–24.

Owen, R., Stilgoe, J., Macnaghten, P., Gorman, M., Fisher, E., & Guston, D. (2013). A Framework for Responsible Innovation. In R. Owen, M. Heintz, & J. Bessant (Eds.): *Responsible innovation* (p. 27–50). Chichester, UK: Wiley.

Rice, W. R., Friberg, U., & Gavrilets, S. (2012). Homosexuality as a consequence of epigenetically canalized sexual development. *The Quarterly Review of Biology, 87*(4), 343–368.

Rip, A., & Voß, J.-P. (2013). Umbrella terms as a conduit in the governance of emerging science and technology. *Science, Technology and Innovation Studies, 9*, 39–59.

Rothstein, M. A., Cai, Y., & Marchant, G. E. (2009). The ghost in our genes: Legal and ethical implications of epigenetics. *Health Matrix, 19*(1), 1–62.

Rothstein, M. (2013). Legal and ethical implications of epigenetics. In R. L. Jirtle & F. L. Tyson (Eds.), *Environmental epigenomics in health and disease* (p. 297–308). Berlin/Heidelberg: Springer.

Smith, L. C., Suzuki, J., Goff, A. K., Filion, F., Therrien, J., Murphy, B. D., et al. (2012). Developmental and epigenetic anomalies in cloned cattle. *Reproduction in Domestic Animals, 47*, 107–114.

Star, S. L., & Griesemer, J. R. (1989). Institutional ecology, 'translations' and boundary ob-jects: Amateurs and professionals in Berkeley's Museum of Vertebrate Zoology, 1907–39. *Social Studies of Science, 19*(3), 387–420.

van Montfoort, A. P., Hanssen, L. L., de Sutter, P., Viville, S., Geraedts, J. P., & de Boer, P. (2012). Assisted reproduction treatment and epigenetic inheritance. *Human Reproduction Update, 18*(2), 171–197.

von Schomberg, R. (2013). A vision of responsible innovation. In Owen, R., Heintz, M., & Bessant, J. (Eds.): *Responsible innovation. Chichester* (p. 51–74). Chichester, UK: Wiley.

Weinhold, B. (2006). Epigenetics: The science of change. *Environmental Health Perspectives, 114*(3), A160–A167.

Youngson, N. A., & Whitelaw, E. (2008). Transgenerational epigenetic effects. *Annual Review of Genomics and Human Genetics, 9*(1), 233–257.

Author Biographies

Reinhard Heil M.A. is a researcher at the Institute for Technology Assessment and Systems Analysis (ITAS) at the Karlsruhe Institute of Technology (KIT). His areas of interest include big data, life sciences (synthetic biology, epigenetics) and transhumanism. Some of his publications (selection) are as follows: Frank, D.; Heil, R.; Coenen, Chr.; König, H. (2015): Synthetic biology's self-fulfilling prophecy—dangers of confinement from within and outside. Biotechnology Journal 10 10 (2), S. 231–235, publ. online, DOI: 10.1002/biot. 201400477; König, H.; Frank, D.; Heil, R.; Coenen, Chr. (2013): Synthetic genomics and synthetic biology applications between hopes and concerns. Current Genomics 14(2013)1, S. 11–24, DOI: 10.2174/1389202911314010003.

Contact: KIT—The Research University in the Helmholtz Association, Institute for Technology Assessment and Systems Analysis (ITAS), Karlstraße 11, D-76133 Karlsruhe.

Stefanie B. Seitz Dr. rer. nat. was senior scientist in the research area innovation processes and impacts of technology of the Institute of Technology Assessment and Systems Analysis (ITAS) at the Karlsruhe Institute of Technology (KIT) until 01/2016. She has published about the governance of manufactured nanomaterials, epigenetics and synthetic biology. Therefore, her research is focused on the question how society deals with the dilemma that new and emerging sciences and technologies poses and how the big promises can materialize and precaution with respect to uncertainties and potential risks can be maintained at the same. Moreover, she is interested in the potentials of public participation and the concepts of responsible research and innovation.

Contact: DBFZ – Deutsches Biomasseforschungszentrum gGmbH, Torgauer Str. 116, 04347 Leipzig, Germany.

Jürgen Robienski Dr. rer. publ. is a German lawyer in Hannover and Müden/Aller (Lower Saxony). He is a research fellow at the Center of Ethics and Law in the Life Sciences (CELLS) of Leibniz University in Hannover. Some of his publications include: „Die Auswirkungen von Gewebegesetz und Gendiagnostikgesetz auf die biomedizinische

Forschung – Biobanken, Körpermaterialien, Gendiagnostik und Gendoping" Hamburg (2010), Verlag Dr. Kovac; „Ethische und rechtliche Aspekte im Umgang mit genetischen Zufallsbefunden, Herausforderungen und Lösungsansätze", (with Rudnik-Schöneborn, S., Langanke, M., Erdmann, P.), in: Ethik in der Medizin 2013, DOI 10.1007/s00481-013-0244-x; „Aktuelle medizinrechtliche und -ethische Herausforderungen der Pathologie" (with: Hoppe, Nils), Der Pathologe 2013 34(1). His topics of research are biomedical law, biotechnical law, labor law, biobanking, life Sciences.
Contact: Eichenkamp 6, 38539 Müden.

Harald König holds a Ph.D. in biology and works since 2011 as a senior researcher at the Institute for Technology Assessment and Systems Analysis (ITAS) of the Karlsruhe Institute of Technology. His research in molecular biology on gene expression and signal transduction pathways as well as his more recent work in technology assessment resulted in various publications in peer-reviewed scientific journals [Signal-dependent regulation of splicing via phosphorylation of Sam68, Nature 420, 691 (2002); Synthetic genomics and synthetic biology applications between hopes and concerns, Current Genomics 14, 11 (2013); Synthetic biology's self-fulfilling prophecy—dangers of confinement from within and outside, Biotechnology Journal DOI: 10.1002/biot.201400477 (2015)]. His research interests focus on societal challenges from the life sciences and biotechnology and on governance and science policy issues.
Contact: KIT—The Research University in the Helmholtz Association, Institute for Technology Assessment and Systems Analysis (ITAS), Karlstraße 11, D-76133 Karlsruhe.

Introduction to Epigenetics

Jörn Walter and Anja Hümpel

Abstract

Epigenetic processes control central genomic functions such as the utilization of genetic information over the course of life. Epigenetic processes are controlled by adding and removing epigenetic modifications on the genes. Epigenetic modifications are added at different molecular levels and form a complex combination of positively and negatively regulating molecular signals. Most of these signals are established directly on the DNA bases or on the proteins that package the DNA, called histones. Modern sequencing methods make it possible to locate these various types of epigenetic modification with precision and to associate their functional significance with a particular gene-specific control. Epigenetic modifications are cell-specific, and their function must therefore be viewed and evaluated in a different way to genetic changes, which are the same in all cells. In epigenetic studies, therefore—unlike genetic analysis—the cell type or (in tissues) the cell composition must always be included in the picture. Cell-type-specific epigenetic patterns can be affected by factors that are endogenous to the organism (ageing, hormonal control) and by those that are exogenous (environment, e.g., metabolism, stress), and they lead

J. Walter (✉)
FR 8.3 Biowissenschaften, Genetik/Epigenetik,
Naturwissenschaftlich-Technische Fakultät III, Universität des Saarlandes,
Postfach 151150, 66041 Saarbrücken, Germany
e-mail: j.walter@mx.uni-saarland.de

A. Hümpel
Interdisziplinäre Arbeitsgruppe Gentechnologiebericht, Berlin-Brandenburgische
Akademie der Wissenschaften, Jägerstr. 22/23, 10117 Berlin, Germany
e-mail: huempel@bbaw.de

© Springer Fachmedien Wiesbaden GmbH 2017
R. Heil et al. (eds.), *Epigenetics*, Technikzukünfte, Wissenschaft und
Gesellschaft / Futures of Technology, Science and Society,
DOI 10.1007/978-3-658-14460-9_2

to persistent changes in cell programming. As a general principle, cell-type-specific epigenetic differences are considerably more stable and more pronounced than changes arising due to exogenous factors. Epigenetic modifications are stably passed on through cell divisions. However, when cell programming changes, they are deleted or their composition is altered (reprogrammed). In human beings, large-scale reprogramming (deletion) of old 'inherited' epigenetic modifications takes place both in gametes and in the embryo shortly after fertilization. For this reason, transmission of 'acquired' epigenetic modifications across generations is possible only to a very limited extent in humans.

1 Basic Principles of Epigenetic Concepts

The word 'epigenetics' means roughly 'above genetics', with undertones of 'in addition to the genome' (see Seitz as well as Schoul, both in this volume). Epigenetics describes mechanisms that lead to changed, heritable structural and activation states of the chromatin[1] without changes to the primary nucleotide sequence[2] (definition by Knippers and Nordheim 2015; see Chap. 20, this volume, for an overview of the molecular aspects). This molecular genetic definition describes some characteristics, but leaves out some aspects of the functional consequences and other possible levels of epigenetic control. In this contribution, therefore, we shall briefly explore some of these additional aspects of epigenetics.

---

[1]Chromatin is the name given to the entirety of the DNA and protein material in the cell that is stained by alkaline staining. Core elements of chromatin are histone protein complexes, which together with the DNA wrapped around them form the nucleosomes. Almost all the DNA in the chromosomes is organized ('packaged') in nucleosomes. Between nucleosomes lie short regions of free (non-coding) DNA ('spacers'). Chromosome regions that play a part in gene regulation are less densely covered in nucleosomes. Nucleosomes are differently arranged in active and in inactive gene regions; they can also be even more tightly 'packed' in higher-order structures. Such higher-order structures are usually completely inaccessible to gene regulation. Other material that occurs in chromatin includes site-specific RNA molecules and other proteins that are not histones but are important for gene regulation, or control gene regulation in a targeted way.

[2]'Nucleotide sequence' means the sequence of chemical building blocks of DNA (and RNA). DNA forms long chains of linear molecules in which nucleobases and pentoses (sugar) are linked together by phosphates. These molecules are copied by enzymes, and thus the information carried by the molecules is duplicated and transmitted onwards.

The aim of this brief digression is to roughly outline the range of current epigenetic concepts and the differences between them.

Whereas definitions of genetics-oriented epigenetics focus primarily on the aspect of direct heritability via modified DNA bases and chromatin modifications, definitions of epigenetics that are geared more to cell biology, or those that are more purely operational, see it more as a portmanteau word for mechanisms additional to DNA that induce heritable alterations in cellular programming and that can also take place at other levels than DNA. One limitation of definitions based on strict heritability of epigenetic modifications is that even 'classical' epigenetic modifications do not act solely to enable transmission of epigenetic states of DNA and chromosomes: they often also influence other forms of temporary regulation of genomes, such as DNA replication, DNA recombination, short-term base changes (mutations), DNA repair, and transient (non-heritable) gene control. These temporary processes do not in the strict sense lead to stable, heritable phenotypic changes.

In broader definitions the term 'epigenetics' often serves as a kind of overall designation non-genetic heredity at every level; that is, it describes a number of sometimes very different mechanisms whose temporal and heritable components have not in all cases been clearly determined. For example, the passing on of small RNA molecules ('small RNAs') from cell to cell is regarded as epigenetic transmission—even though this is primarily a temporary genetic effect determined by the cell plasma and does not take place in the cell nucleus. In addition, various processes of RNA storage and RNA interference are often referred to together as 'epigenetic'—for some of them this is an accurate description, but for others it is very hard to argue that it is valid. In other interpretations that go still further, even molecular processes in the cell plasma about which little is known, for example the spatial reconfiguration of prions,[3] are cited as examples of epigenetic phenomena (Lewin 1998). Another aspect is 'early embryonic programming', the process by which, under certain circumstances, proteins and RNA molecules passed on with the cell cytoplasm of oocytes and spermatozoa can affect gene expression in the long term. In animal and plant breeding, examples of this are known in the form of (reciprocal) hybrid crossings with differentially strongly expressed traits (Youngson and Whitelaw 2008).

---

[3]Prions are small glycoproteins (that is, proteins that contain sugar chains), whose physiological function is still unclear. This class of protein became known through pathological (disease-causing) such as those that cause Kreutzfeld–Jakob disease or 'mad cow disease' (bovine spongiform encephalopathy).

The epigenetic processes referred to above make clear how complex are the regulatory aspects that must be considered in relation to the specific biological context in any individual case. In fact, no single, generally valid definition of epigenetics can cover the multiplicity of mechanisms known to science today, some of which go beyond the purely genetic level.

Accordingly, it remains true that, even in the specialist literature, 'epigenetics' continues to be a broadly used term that in many cases inadequately reflects to the systemic processes behind it.

2   Levels of Epigenetic Gene Control

As we have said, epigenetic mechanisms are located at several levels. On the genome, the levels are those of DNA modifications and chromatin. Partly decoupled from the genome there are modifying proteins and non-coding RNA, whose site of action is in the nucleus or the cytoplasm. The common property of all three of these levels of epigenetic mechanism is that they influence the function and regulation of genes in a *long-lasting* but at the same time *reversible* manner.

2.1   DNA Methylation

DNA methylation is added by DNA methyl transferases (DNMTs) to certain building blocks of DNA (bases) in a targeted way. DNA methylation is a chemically very stable covalent modification of certain cytosine bases which can be (indirectly) demonstrated in old DNA. Through its attachment to DNA bases, it serves as a direct signal for a copying procedure performed by the DNA methyltransferase DNMT1 after DNA replication. In this way, DNA methylation can be directly copied and passed on through cell divisions. In a similar way to histone modifications, the cells of our body cell show specific DNA methylation patterns. In the early stages of development, the amount of DNA methylation in the genome is very strongly reduced. Then in the course of development it is restarted in a cell-specific way, and during this process DNA methylation is started in the genome in a very targeted manner. Cytosine building blocks are mostly methylated in the sequence cytosine–guanine (CpG). In neurons and stem cells, methylation is also widely found outside CpGs, but the functional of this "irregular" non-CpG-methylation is still unclear. DNA methylation is recognized as an epigenetic signal by special DNA-binding proteins which translate the epigenetic signal into a function. Depending on its position (site and methylation status), DNA methylation acts as a repressing (often) or activating (less often) epigenetic signal. In large parts of the genome, DNA methylation serves as a signal to inactivate repetitive DNA structures and 'jumping genes' (transposons). In addition, a number of genes are switched off long-term by DNA methylation.

DNA methylation exists in almost all multicellular organisms except for the classical model organisms in developmental biology *Drosophila melanogaster* (fruit fly) and *Caenorhabditis elegans* (roundworm). In all the organisms in which it is found, DNA methylation has a gene-regulating function. Insects (bees, termites, ants) have highly developed systems for DNA methylation and also for histone modification. Their purpose is to control genes that are important for morphological changes during reproduction, but it is likely that they also have a part in controlling learned and adaptive behaviours (Wang et al. 2006; Maleszka 2008). Observations in bees, for instance, have shown that queens and various other workers differ epigenetically, that queen differentiation is caused by epigenetic changes due to nutritional substances.

In plants, too, DNA methylation plays an essential epigenetic role (Henderson and Jacobsen 2007). A number of heritable, adaptive epigenetic effects are seen in plants that rely on DNA methylation (Hirsch et al. 2012). Plants possess a very highly developed enzymatic system of control over DNA methylation, and some very specialized forms of epigenetic regulation may be seen. It was in plants that researchers showed for the first time that DNA methylation can be actively removed by DNA repair processes (Zheng et al. 2008). Later, similar mechanisms were demonstrated in some vertebrates (zebrafish and *Xenopus*) and in mammals (mouse and human) (Gehring et al. 2009).

In mammals, including humans, DNA methylation can occur in other forms of modification—with highest abundance in stem cells and neurons. Building on 5-methylcytosine, additional modifications catalysed by the Tet enzymes occur in three oxidation states: 5-hydroxy-methylcytosine, 5-formyl-cytosine, and 5-carboxy-cytosine. 5-Hydroxy-methylcytosine (5OH-cytosine) is recognized by special proteins and interpreted differently to simple DNA methylation (e.g., not correctly copied during replication). The higher oxidation steps 5-fluorocytosine and 5-carboxycytosine most likely serve as recognition signals for DNA repair; that is, they are only short-lived and are then removed again from the DNA. There are clear indications that oxidative modifications are important for loss of DNA methylation in early germ cell and embryonic development (Wossidlo et al. 2011; Seisenberger et al. 2013; Arand et al. 2015). The significance of the oxidative forms of DNA methylation that occur especially in gametes, early embryos, stem cells and neurons is currently being investigated in detail. In all these cells types, wide-ranging epigenetic changes are observed during the course of development (stem cells) and ageing (neurons). This suggests that the various forms of DNA methylation in these cells are used for short-term switches in gene programming. The most recent studies in stem cells indicate that DNA methylation adapts to rapid and extreme changes in culture conditions (e.g. the culture media), and that oxidative modifications play a part in this (Ficz et al. 2013; Habibi et al. 2013; Azad et al. 2013). Similar processes may also be in play in other environmentally induced alterations in somatic cells (e.g. neurons).

2.2   Histone Modifications

The DNA of our genome is packaged in nucleosomes. Nucleosomes are made up of eight histones wound around the roughly 150 bases (the building blocks of DNA). DNA packaged in nucleosomes is not directly accessible to biochemical processes such as gene transcription. Nucleosomes are therefore distributed in a gene-specific way. The histone protein modifications are very important to the strength of the packaging and to nucleosome distribution. These modifications control a series of epigenetic processes (Kubicek et al. 2006). The main entities modified are particular amino acids in the start and end regions ("tails") of histone proteins H3 and H4. Histone modifications are extremely rich in variants: at the present time, a total of 140 histone modifications are known. They are always post-translational modifications,[4] usually of basic polar amino acids such as serine, threonine, lysine and arginine. The modifications are small chemical functional changes in the form of acetylation, methylation, phosphorylation, ubiquitination and SUMOylation (for an overview, see Kouzarides 2007). In functional terms, a distinction can be made between chromatin-opening and chromatin-closing modifications, which respectively promote or inhibit the reading of genes.

Histone modifications are added to histones when the latter are in nucleosomes, at particular locations in the chromosome. The modifications are brought about by special enzymes. Locating and 'deciding' exactly which nucleosomes are to be modified, and in what way they are to be modified, is achieved with the help of other enzymes/proteins (e.g. transcription factors) which position the histone-modifying enzymes in the 'right' place in the chromatin (nucleosome). Histone modification in some cell types follows a very specific sequence and combination on the nucleosomes along the chromosomes. They thus 'encode' the nucleosome packaging, determining which genes are switched on or off. They also mark the regulation sites for the reading (transcription) of the genes and determine

---

[4]Post-translational modifications are modifications carried out on the 'mature' protein, after translation—of the nucleotide sequence into an amino acid sequence—has been completed.

the speed at which transcription takes place. Other entities important for turning epigenetic modifications into genetic activity are enzymes that can relocate nucleosomes within chromosomes. They are needed in order to release gene-controlling elements from the nucleosomes. To do this, the enzymes read and interpret the local histone modification structure of the nucleosomes. Among others, the so-called polycomb group protein complexes and their antagonists the trithorax complexes are responsible for the exact formation of cell-type-specific histone modifications on the nucleosomes along the gene. These complexes position histone modifications precisely on regulatory segments of the genome and mark these as switched on or off (Whitcomb et al. 2007). Histone-demodifying enzymes can remove these modifications locally or transform them and can thus reverse processes. It is not just the switching on and off of genes that is regulated in this way: RNA splicing and maturation, DNA replication, and DNA repair are all influenced or controlled via histone modifications (Corpet and Almouzni 2009; Varga-Weisz and Becker 2006).

In the course of cell development, gene-specific histone modifications are added and removed in a precisely ordered sequence of events. During development, cell-specific patterns of histone modifications are established in this way, step by step, in every type of cell in the body. Every cell type possesses, in parallel to the DNA methylation described above, its own 'epigenomic signature'. In stem cells, a very specific 'immature' double combination of activating and repressing histone modifications is observed in gene-regulatory regions, allowing these cells to maintain an intermediate epigenetic status which is essential to their ability to retain pluripotency (that is, the quality of stem cells that allows them to differentiate into any kind of cell) (Bernstein et al. 2006; Mikkelsen et al. 2007; Chi and Bernstein 2009). These gene-regulatory regions, which are in an epigenetically neutral waiting condition, react extremely swiftly to exogenous differentiation stimuli; that is, they can quickly switch epigenetically in order to carry out the special tasks of a differentiated cell. Later on in the course of differentiation, increasingly large regions of the genome are then marked by particular histone modifications in such a way that they are permanently closed and only those genes that are necessary to the cell remain switched on.

The modifications of histones can be analysed along the chromosomes using a technique called chromatin immunoprecipitation (ChIP). This involves enriching the modified histone within the nucleosomes with the help of antibodies, each of which binds selectively to one type (and one type only) of modified histone. The DNA of the bound nucleosome fractions is then isolated. High-throughput sequencing of this DNA (ChIP-Seq) allows the determination of the parts of the genome where the nucleosomes with the identified histone modifications were bound. Since the genome sequence is known, the histone modifications can be 'mapped' along the DNA. These histone modification maps demonstrate that gene-regulatory regions, gene-encoding regions, and segments that lie between the genes, differ markedly in their histone modification patterns. Mapping seven to eight histone modifications appears to be enough to divide up the genome functionally into segments of active genes and gene switches and regions of inactive genes. Together with DNA methylation maps and gene expression data, this provides a wealth of gene- and cell-specific information giving insight into both healthy and diseased cells (Karnik and Meissner 2013). Special electronic mapping aids (known as epigenome browsers) make it possible to analyse complex datasets together and make use of them for gene-specific research. Many examples now exist of how histone modification mapping has provided clues towards a gene-specific, functional interpretation of the molecular causes of disease. Accurate histone modification mapping would therefore provide direct access to the understanding of disease. However ChIP-Seq technologies are of only limited usefulness for diagnostic mass screenings, since it requires large quantities of fresh cells (up to 10^6) for its performance, and the investigation must also be carried out under extremely standardized conditions for sample comparisons to be valid. In the worldwide coordinated analysis of many isolated cell types,[5] it is becoming increasingly clear that some epigenetic information content can be obtained even from the DNA methylation. The patterns of DNA methylation in part follows the distribution of histone modification pattern on histones. Since DNA can be harvested from almost all cells (including frozen ones) in adequate quantities, most epigenomic diagnoses focus on changes in the DNA methylation. The data obtained in this way are subsequently interpreted for functional information using comparisons to histone reference patterns.

All eukaryotes display histone modifications. A large number of enzymes are responsible for adding, removing, and actually recognizing them. Interestingly, species-specific differences occur: the effect of a histone modification or added patterns can vary from organism to organism. Also differences in the molecular interplay between DNA level and histone level are seen. In the extreme case, one level of epigenetic control can even be absent altogether, as is shown by the absence of any DNA methylation in *C. elegans* and several other organisms.

Histone modifications are established at the protein level; that is, they are not located directly on the DNA. Histones are components of nucleosomes, which are duplicated at every cell division and have to be redistributed onto the replicated

---

[5]See data from the International Human Epigenome Consortium (IHEC): http://ihec-epigenomes.org/.

DNA. How histone modification patterns can be stably inherited on the gene locus during the process of replication (cell division) is a question that has not yet been definitively answered. Some early models and molecular clues exist; these show that nucleosomes remain on their gene locus despite replication and cell division, and that the epigenetic information of the 'old' nucleosome is transferred to the new nucleosome by a kind of copying procedure (see Knippers and Nordheim 2015).

2.3   Epigenetics and 'Non-coding' RNAs

One prominent example of expanded epigenetic control mechanisms is the transcriptional and post-transcriptional "switching-off" or silencing of genes/transcripts by small regulatory RNAs. These epigenetic regulatory mechanisms were first discovered in plants. The subsequent discoveries of RNA-mediated epigenetic regulatory phenomena in almost all organisms—including human beings—increasingly show the close link between RNA-mediated regulatory processes and epigenetic transmission. In particular, small RNAs, such as pirRNAs (in gametes), miRNAs and siRNAs (in all cells), and long non-coding RNAs (lncRNAs) play an important part in establishing or implementing epigenetic processes. RNA is very diverse, not only in its structure, but also in its function. A general distinction should be made here between direct epigenetic effects, indirect intermediary functions, and subsequent implementation functions of non-coding RNAs.

Close interplay between structural and catalytic RNAs and epigenetic modifications is characteristic of many model organisms (yeast, *Drosophila*, *C. elegans*, *M. musculus*, *Arabidopsis thaliana*), but the significance of small RNAs was first identified in connection with expression control and chromatin structures in plants in particular. To put it another way: our conceptual understanding of how small RNAs come into existence and how they operate originates from plant epigenetics (Baulcombe 2004).

The expression of these small RNAs is often regulated in a cell-specific way and controlled via epigenetic modifications (e.g. via promoter methylation or chromatin modifications). Small RNAs have considerable influence on the translation and stability of mRNAs.[6] In addition to that, they assume an important

---

[6]Messenger RNA (mRNA) is the complementary copy of a coding gene sequence of DNA. This copy transports the information about the gene out of the cell nucleus into the cytoplasm, where it serves as a matrix for biosynthesis of a protein; that is, it is translated into the amino acid sequence of a protein.

function in controlling the formation of heterochromatin by leading histone-modifying and DNA-modifying enzymes to particular target regions such as the centromeres and telomeres, as well as transposable elements.[7] These mechanisms of effect are assumed primarily by special classes of small (si, casi, pi) RNAs. In human beings, close interplay can be demonstrated between the small dsRNAs and gene regulation in the epigenetically controlled (through DNA and histone modification) promoter control of the ribosomal gene cluster.[8] The same is true of imprinting regions.[9] Because of this, piRNAs are centrally important to epigenetic control of gamete development. They lay the foundation for epigenetic silencing of transposable elements (jumping genes) in maturing gametes. Besides the small RNAs, crucial roles in the epigenetic control of gene activity are similarly played by long non-coding RNAs (lincRNAs) such as XIST or AIR and HOTAIR. It turns out that the lincRNA XIST is essential for the silencing of genes on the X-chromosome (gene dosage compensation) in human beings (Clerc and Avner 2006), and in the process it regulates the stable, long-term formation of certain alterations in histone and DNA methylation.

One of the things that can be found in cancer cells is epigenetically misregulated expression of miRNA host transcripts, which results in incorrect regulation of genes targeted by the miRNA. Similar processes are seen in plants and in single-celled organisms (*Paramecium*). This observation requires that the direct association between the expression of small RNAs and other epigenetic gene regulatory cascades should be tested. The functional significance of this observation ranges from direct gene control in the course of development to defence against viruses (inactivation).

At present it is not possible to judge the importance of small and long non-coding RNAs in the control of epigenetic processes in human beings. This is partly because, for reasons that are still incompletely understood, these RNAs occur in such a multiplicity of forms. It is also because their interactions with epigenetic controls at other levels are very diverse. Recent findings show, for example, that a hitherto relatively unknown type of long circular RNAs (circRNAs) play a significant part in how the small RNAs function (circRNAs serve among other things as miRNA 'stores'or 'sponges'). Based on these effects, which

---

[7]In contrast to euchromatin, heterochromatin is a densely packed, inactive chromatin.

[8]Ribosomal gene clusters are clusters of genes that code for components of ribosomes, the cell's 'translation apparatus'.

[9]Imprinting is the transmission from one generation to the next of DNA and histone modifications that result in silencing of the marked ('imprinted') copies of genes (alleles) from one parent, leading to preferential expression of the alleles from the other parent.

are observed in a very wide variety of model organisms, it may at least be assumed that in humans, too, there is a close interdependence between small RNAs that have structural and enzymatic effects and epigenetic control of the functions of the genome. This is one reason why, in terms of research policy, it will be of absolutely fundamental importance to link together and promote research in these areas, which need to be brought closer together.

3 Epigenomics

Epigenomics (research into the epigenetics of the genome) is one of the youngest branches of epigenetics. The goal of epigenomics is to read, locate, and interpret the complete set of the various levels of epigenetic control of the entire genome. In addition, it aims to compare epigenetic maps between cells and draw conclusions from them about development, disease and ageing.

During development hundreds of special cell types with millions of descendent cells are generated. In this process cells of a given cell type acquire a characteristic epigenetic program. This is referred to as the epigenome (Bernstein et al. 2010). The epigenome of a cell is a direct reflection of the gene activation state of the cell. It encodes the information about how and where gene specific activation switches are located and used in the genome (ENCODE Project Consortium et al. 2012). Epigenome analyses are complex as they combine the layers of DNA methylation patterns, histone modifications and RNA expression. All layers must always be looked at together: that is, the genome needs to be 'scanned' for each modification separately and later combined into the epigenome. Next-generation DNA sequencing techniques have opened up hitherto unknown possibilities to achieve this, enabling the creation of high-resolution epigenome maps of normal and diseased cells. Epigenomics requires complex bioinformatics software for its visualisation and interpretation. The boom in epigenomics research in recent years was triggered by extensive national and international programmes, all united under the auspices of the International Human Epigenome Consortium (IHEC). Even preliminary results testify to the enormous importance of this research direction, which is opening up deep new vistas, hitherto unknown to us, into the basic epigenetic patterns of healthy and diseased cells.

The high-resolution epigenetic mapping techniques allow genetic and epigenetic changes to be analysed simultaneously. In comparative studies, it is increasingly observed that the changes in DNA methylation and in chromatin modifications often co-occur with small genetic variations (base exchanges, insertions, deletions)—that is, that genetics and epigenetics are in fact very often closely interwined.

Epigenomic data provide deep cell-specific information: they explain how the unique genetic programs take effect and is translated in hundreds of cell specific programs. In addition epigenomic data are a rich sourece to explain the function of genetic variants in the human genome. Over the past 20 years, a large number of genome-wide genetic association studies (GWAS) have been carried out for many common diseases—often with somewhat disappointing results. Making use of epigenetic data (DNA methylation and chromatin accessibility) enables a completely new evaluation of these genetic data and offers new functional explanations that make sense of the genetic data.

Epigenomic data convey very deep information about the genetic and epigenetic predispositions of an individual. For this reason, these data need to be handled with great care and discretion, and questions about the ethical and legal aspects of privacy relating to the use of these epigenetic data need to be debated. We do not yet know how to estimate (because of the lack of case numbers and comparators) whether epigenomic data contain permanent traces of personal epigenetic adaptations to life circumstances and thus to lifestyle (e.g. drug abuse, smoking, etc.). It is clear that personal age and cell age can be read from the epigenome. Epigenome mapping will open up new areas of personal diagnosis, and will provide answers to the intensely debated questions of how far the environment leaves long term marks on and influences the function of our genes. In epidemiology, the number of genome-wide epigenetic association studies has therefore increased enormously.

The technical means required for epigenetic diagnosis of personality features already exist. However, our ability to interpret the mass of data is still extremely limited. The complexity of the data creates a mass of sometimes contradictory possible interpretations which require complex computer processing before they can be turned into medically meaningful statements. However, besides these obstacles it is already clear that a personal genetic diagnostic of the future will include epigenetic data. DNA methylation is the level element to analyse. It can be read relatively easily and reliably, genome-wide, using high-throughput technologies—even from small quantities of cells and frozen material, and even indirectly, 'retrospectively,' in very old DNA (as recently shown by the reading of Neanderthal DNA).

Genome-wide epigenetic studies also play an increasing role in the diagnostics of human diseases. Epigenetic comparisons of cancer types identified cancer specific signatures (Weisenberger 2014). These signatures provide information about the origin and the prognosis of many cancers. While the causes and influence of epigenetic changes on cancer development are not yet well understood it becomes more and more clear that the restructuring of genome wide epigenetic program is a hallmark of cancer, causing an uncontrolled biological development. This feature is accompanied in many cancers by the frequent occurance of mutations in enzymes controlling epigenetic programs. Hence both genetic and epigenetic changes must be taken into consideration for a full understanding of cancer biology and cancer therapies. At present methylation patterns are being increasingly used for development of early cancer recognition tests in body fluids. The first diagnostic products already achieved the leap into clinical use or are at the point of being approved for routine testing.

4   Lessons from Epigenomics

All epigenetic mechanisms serve organisms primarily for differentiated use and control of genetic information. Basic epigenetic mechanisms are found throughout the animal and plant realms, but there they are used (in a way adapted to each organism) for long-term control of genes in the course of development and ageing. In higher organisms and in human beings, the various levels of epigenetic regulation become increasingly complex and specific. Thus, epigenetic mechanisms always need to be viewed in the context of the organism under investigation and of its genome and the research question to be answered. Caution is needed when drawing conclusions from epigenetic processes observed in model organisms and applying them to the human case, and good evidence is mandatory. This is particularly the case for phenomena relating to transgenerational transmission of epigenetically altered gene status or long-term adaptation to changed environmental conditions.

4.1   Epigenetics and Adaptation

Reaction of an individual's genes to environmental stimuli and lifestyle is not a new observation. The individual's basic genetic apparatus also offers a different framework of responses to environmental stimuli. We now often speak in terms of 'epigenetic adaptation'. This term is also associated with neo-Lamarckism and suggests a form of individually programmed adaptation of the organism to altered living conditions. Two elements are often left out of account here. Firstly, epigenetic processes primarily serve to control development and the maintenance of vital functions (health and ageing); that is, epigenetically controlled developmental processes are genetically determined and can only vary within certain limits.

Secondly, there are few indications that environmental influences directly produce targeted epigenetic variation of individual cell programs, rather than that the primary reaction to the environment triggers a secondary epigenetic reaction.

Irrespective of the causality question, the extent of epigenetic adaptability will depend primarily on individual genetic configuration and variation. What is known today indicates that epigenetic processes modulate the genetic range of play—but no new levels of regulation come into existence. Hence, it is always important to ask whether the cause of an observed epigenetic change is to be found at a level above the gene sequence, or whether it is actually coupled to gene variants.

Thus, epigenetic control must be viewed not only as a matter of the switching on or off of genes, but, for many examples of individual variation, as a process of limiting the modulability of genetic information. Epigenetic modifications in effect determine the framework within which genetic information is used. In consequence, it is important to view epigenetic phenomena from a quantitative biological viewpoint.

There are a series of inherited phenomena that are determined epigenetically, such as the part-of-origin imprinting of genes (genomic imprinting) or the silencing of one of the two X chromosomes in women. In both phenomena, the development of the organism is coupled to precisely regulated, fixed epigenetic control in an obligatory manner. Accordingly, epigenetic impairments arising in connection with imprinting and X inactivation lead to serious biological consequences such as syndromal diseases.

4.2   Concepts of Epigenetic Inheritance in Humans

A fundamental characteristic of epigenetics is its heritability; that is, stable transmission or handing on of fixed epigenetic markings that survive cell divisions. In contrast to true mutations, however, epigenetic modifications ('epimutations') can be reversed and can be deleted in a targeted way. The heritability of epigenetic modifications (histone modifications and DNA methylation) beyond cell divisions is without doubt a core characteristic of all multicellular organisms. Transmission via the germ line and haploid gametes, on the other hand, cannot be assumed with certainty for all organisms. Except for genomic imprinting, epigenetic modifications in the parent generation in humans are only very sporadically passed on to the children. In much-cited examples of early epigenetics research, the observed epigenetic changes are at least partially coupled to genetic parameters (transmission via the cytoplasm) or even genetic changes (changes in the genome). Despite this, such examples are always being used to develop new concepts of the heritability of epigenetic adaptation to the environment from one generation to the next. However, neo-Lamarckist adaptive 'epimutation' scenarios of this kind often prove on closer inspection to rely on very sparse data.

With the exception of genomic imprinting, there are so far no clear indications in man for regularly inherited epigenetic effects through the germ line (Heard and Martienssen 2014). Many observations and reports of transgenerational inheritance of metabolic and stress related phenotypes rely on epigenetic interpretations of empirical data collections (health statistics). The reason for a low abundance of transgenerational heritability in humans is probably the extensive epigenetic reprogramming that occurs in the germ line and after fertilization (in early embryogenesis). Spontaneous occurrence of epigenetic errors that cause diseases, such as erroneous deletion of genomic imprints, cannot really be cited as examples of transgenerational inheritance—because they will not occur anew in the next generation. Frequently cited examples of transgenerational effects in animal models include the nutritionally determined (folic acid) change in the 'viable yellow' (fur colour) gene in agouti mice,[10] but closer scrutiny shows that here epigenetic programs are tightly coupled to a genetic mutation and the genetic background of the animals (Whitelaw and Whitelaw 2006). A recent report on a clear inheritance of a metabolic phenotypes in female and male mice strongly points in the direction of epigenetic changes but the exact molecular mechanisms remain unresolved (Huypens et al 2016).

About one fact there is no debate: that early epigenetic modulation of the (inherited) parental genomes by factors from the maternal oocyte cytoplasm can have a long-term effect on the individual epigenetic expression of genes. The presence of small RNAs and certain modifications of proteins that come into contact with the parental chromosomes via the oocyte cytoplasm can have a lasting influence on gene regulation.

In plants, on the other hand, the indications regarding transgenerational epigenetic inheritance can be interpreted much more clearly. Epigenetic inheritance from one generation to another is regarded as confirmed for some plant phenotypes/genes. Some of these phenomena have also been demonstrated at the molecular level (Henderson and Jacobsen 2007). In plants, unlike in animals,

---

[10]Agouti mice carry a special variant, named 'agouti viable yellow' (Avy), of a gene that determines hair colour. The more intensely methylated this gene is, the darker the hair—and the healthier the mouse. Supplementing the feed of agouti mothers with methylating molecules such as methionine, folic acid and zinc results in more intensely methylated Avy genes in the offspring, even the grandchildren. This experiment is often used as an example of the epigenetically mediated influence of lifestyle on the health of later generations.

deletion of epigenetic modifications in the gametes does not take place. Some acquired epigenetic changes can be retained for generations. Two hundred and fifty years ago Linnaeus and Goethe described a mutant snapdragon (altered flower form) that finally proved to differ from its nearest relative by just one epimutation (Cubas et al. 1999).

4.3   Perspectives in Epigenetic Research

Epigenetics has already made its way into applied biomedicine at many levels. Epigenetic processes play a preeminent role in the production of synthetic and natural stem cells. Epigenetic analysis is increasingly being used to generate personal biomarkers, e.g. for the differential diagnosis and early recognition of cancer. In addition, epigenetic processes offer a way to approach the development of new kinds of epigenetic drug substances. Some of these substances directed against DNA and histone modifications are already being successfully used to treat specific cancers. Many others substances directed against epigenetic modifier enzymes are currently under development.

In preventive health care, psychology and the social sciences, epigenetic mechanisms are already being debated as possible factors influencing personality. This debate, however, rests on very few concrete data. Repeatedly, a few examples, mostly from model organisms, are used to build up lines of argument that relate to empirically derived biological data such as the Dutch famine study or the Överkalix study. The molecular data relating to these studies, however, are either lacking or can only be assessed to a very limited extent. This is the case for a number of empirical studies in which the epigenetic methods employed often do not meet current standards, or the data have been rather boldly interpreted. The molecular changes observed are often very small and, moreover, are usually statistically overstated. Future comparative studies should be based on more solid methodological and statistical foundations.

As a general principle, epigenetic data and their interpretation require very careful handling. It is entirely possible that epigenetic data reflect information about a person's lifestyle. Epigenomic data should therefore be interpreted and evaluated with care in order to prevent stigmatization.

Epigenetics and epigenetic concepts need to be accorded greater value in the current discourse on human biological questions in the natural and social sciences. It is important to keep a sharp eye out for the basic foundations of epigenetic data and the theories derived from them.

References

Arand, J., Wossidlo, M., Lepikhov, K., Peat, J. R., Reik, W., & Walter, J. (2015). Selective impairment of methylation maintenance is the major cause of DNA methylation reprogramming in the early embryo. *Epigenetics Chromatin, 8*(1), 1.

Azad, N., Rudin, C. M., & Baylin, S. B. (2013). The future of epigenetic therapy in solid tumours – lessons from the past. *Nature Reviews Clinical Oncology, 10*(5), 256–266.

Baulcombe, D. (2004). RNA silencing in plants. *Nature, 431*(7006), 356–363.

Bernstein, B. E., Mikkelsen, T. S., Xie, X. H., Kamal, M., Huebert, D. J., Cuff, J., et al. (2006). A bivalent chromatin structure marks key developmental genes in embryonic stem cells. *Cell, 125*(2), 315–326.

Bernstein, B. E., Stamatoyannopoulos, J. A., Costello, J. F., Ren, B., Milosavljevic, A., Meissner, A., et al. (2010). The NIH roadmap epigenomics mapping consortium. *Nature Biotechnology, 28*(10), 1045–1048.

Chi, A. S., & Bernstein, B. E. (2009). Developmental biology. Pluripotent chromatin state. *Science, 323*(5911), 220–221.

Clerc, P., & Avner, P. (2006). Random X-chromosome inactivation. Skewing lessons for mice and men. *Current Opinion in Genetics & Development 16*(3), 246–253.

Corpet, A., & Almouzni, G. (2009). Making copies of chromatin. The challenge of nucleosomal organization and epigenetic information. *Trends in Cell Biology, 19*(1), 29–34.

Cubas, P., Vincent, C., & Coen, E. (1999). An epigenetic mutation responsible for natural variation in floral symmetry. *Nature, 401*(6749), 157–161.

ENCODE Project Consortium. (2012). An integrated encyclopedia of DNA elements in the human genome. *Nature, 489*(7414), 57–74.

Ficz, G., Hore, T. A., Santos, F., Lee, H. J., Dean, W., Arand, J., et al. (2013). FGF signaling inhibition in ESCs drives rapid genome-wide demethylation to the epigenetic ground state of pluripotency. *Cell Stem Cell, 13*(3), 351–359.

Gehring, M., Reik, W., & Henikoff, S. (2009). DNA demethylation by DNA repair. *Trends in Genetics, 25*(2), 82–90.

Habibi, E., Brinkman, A. B., Arand, J., Kroeze, L. I., Kerstens, H. H. D., Matarese, F., et al. (2013). Whole-genome bisulfite sequencing of two distinct interconvertible DNA methylomes of mouse embryonic stem cells. *Cell Stem Cell, 13*(3), 360–369.

Heard, E., & Martienssen, R. A. (2014). Transgenerational epigenetic inheritance: Myths and mechanisms. *Cell, 157*(1), 95–109.

Henderson, I. R., & Jacobsen, S. E. (2007). Epigenetic inheritance in plants. *Nature, 447* (7143), 418–424.

Hirsch, S., Baumberger, R., & Grossniklaus, U. (2012). Epigenetic variation, inheritance, and selection in plant populations. *Cold Spring Harbor Symposia on Quantitative Biology, 77*, 97–104.

Huypens, P., Sass, S., Wu, M., Dyckhoff, D., Tschöp, M., Theis, F., et al. (2016). Epigenetic germline inheritance of diet-induced obesity and insulin resistance. *Nat Genet, 5*, 497–9.

Karnik, R., & Meissner, A. (2013). Browsing (epi)genomes: A guide to data resources and epigenome browsers for stem cell researchers. *Cell Stem Cell, 13*(1), 14–21.

Knippers, R., & Nordheim, A. (Eds.). (2015). *Molekulare Genetik* (10th ed., p. 568). ThiemeVerlag: Stuttgart.

Kouzarides, T. (2007). Chromatin modifications and their function. *Cell, 128*(4), 693–705.
Kubicek, S., Schotta, G., Lachner, M., Sengupta, R., Kohlmaier, A., Perez-Burgos, L., et al. (2006). The role of histone modifications in epigenetic transitions during normal and perturbed development. *Ernst Schering Research Foundation Workshop 57*, 1–27.
Lewin, B. (1998). The mystique of epigenetics. *Cell, 93*(3), 301–303.
Maleszka, R. (2008). Epigenetic integration of environmental and genomic signals in honey bees. The critical interplay of nutritional, brain and reproductive networks. *Epigenetics, 3* (4), 188–192.
Mikkelsen, T. S., Ku, M., Jaffe, D. B., Issac, B., Lieberman, E., Giannoukos, G., et al. (2007). Genome-wide maps of chromatin state in pluripotent and lineage-committed cells. *Nature, 448*(7153), 553–560.
Seisenberger, S., Peat, J. R., & Reik, W. (2013). Conceptual links between DNA methylation reprogramming in the early embryo and primordial germ cells. *Current Opinion in Cell Biology, 25*(3), 281–288.
Varga-Weisz, P. D., & Becker, P. B. (2006). Regulation of higher-order chromatin structures by nucleosome-remodelling factors. *Current Opinion in Genetics & Development, 16*(2), 151–156.
Wang, Y., Jorda, M., Jones, P. L., Maleszka, R., Ling, X., Robertson, H. M., et al. (2006). Functional CpG methylation system in a social insect. *Science, 314*(5799), 645–647.
Weisenberger, D. J. (2014). Characterizing DNA methylation alterations from the Cancer Genome Atlas. *Journal of Clinical Investigation, 124*(1), 17–23.
Whitcomb, S. J., Basu, A., Allis, C. D., & Bernstein, E. (2007). Polycomb group proteins: An evolutionary perspective. *Trends in Genetics, 23*(10), 494–502.
Whitelaw, N. C., & Whitelaw, E. (2006). How lifetimes shape epigenotype within and across generations. *Human Molecular Genetics, 15*(2), R131–R137.
Wossidlo, M., Nakamura, T., Lepikhov, K., Marques, C. J., Zakhartchenko, V., Boiani, M., et al. (2011). 5-Hydroxymethylcytosine in the mammalian zygote is linked with epigenetic reprogramming. *Nature Communications, 2*, 241.
Youngson, N. A., & Whitelaw, E. (2008). Transgenerational epigenetic effects. *Annual Review of Genomics and Human Genetics, 9*, 233–257.
Zheng, X. W., Pontes, O., Zhu, J. H., Miki, D., Zhang, F., Li, W. X., et al. (2008). ROS3 is an RNA-binding protein required for DNA demethylation in Arabidopsis. *Nature, 455* (7217), 1259–1262.

Author Biographies

Jörn Walter is Professor of Genetics and Epigenetics at the Saarland University in Saarbrücken (http://epigenetik.uni-saarland.de/de/home/). He has a long-standing interest in epigenetic mechanisms focussing on the role of DNA methylation in development and disease. He has contributed to the field through more than 100 research papers in epigenetics and was among the first to discover principles of epigenetic reprogramming in germ cells (P. Hajkova, et al. Mech Dev, 2002. 117 (1–2): p. 15.; Arand, J, et al. Epigenetics & Chromatin, 2015, 8:1 DOI: 10.1186/1756-8935-8-1.) and during early embryogenesis (Oswald et al Current Biology 2002; W. Reik, W. Dean, J. Walter. Science, 2001.

293 (5532): p. 1089–93.; M. Wossidlo, et al. Nat Commun 2011, 2: 241.) His interdisciplinary work bridges between epigenetic and bioformatics (Assenov et al. Nature Methods, 2014, Lutsik et al NAR 2007 human diseases (Souren et al., Genome Biology 2013, 14:R44 doi: 10.1186/gb-2013-14-5-r44.) and work on model organisms. In 1999, he was among the initiators for epigenomic research (Beck et al. Nat Biotechnol. 1999 Dec; 17 (12):1144.) performed the first chromosome-wide epigenetic mapping project (Zhang et al. PLoS Genet 2009 5(3): e1000438.) and currently coordinates the German Epigenome Program DEEP (2012–2017, http://www.deutsches-epigenom-programm.de).

Contact: FR 8.3 Biowissenschaften, Genetik/Epigenetik, Naturwissenschaftlich-Technische Fakultät III, Universität des Saarlandes, Postfach 151150, 66041 Saarbrücken.

Anja Hümpel Dr. is the coordinator and contact person for the topics Synthetic Biology and Epigenetics at the Berlin-Brandenburg Academy of Sciences and Humanities, Interdisciplinary Working Group "Gene Technology Report". Her publications include Gebhard, S.; Hümpel, A.; McLellan, A. D.; Cook, G.M. (2008): The alternative sigma factor SigF of Mycobacterium smegmatis is required for survival of heat shock, acidic pH and oxidative stress. In: Microbiology (Reading, England) 154, 2786–2795. Hümpel, A.; Gebhard, S.; Cook, G.M.; Berney, M. (2010): The SigF regulon in Mycobacterium smegmatis reveals roles in adaptation to stationary phase, heat, and oxidative stress. In: Journal of Bacteriology 192, 2491–2502. Hümpel, A.; Diekämper, J. (2012): Daten zu ausgewählten Indikatoren [zur Synthetischen Biologie]. In: Köchy, K.; Hümpel, A. (Ed.): Synthetische Biologie in Deutschland. Entwicklung einer neuen Ingenieurbiologie? Dornburg (Forum W), 257–285. Köchy, K.; Hümpel, A. (Eds.): Synthetische Biologie in Deutschland. Entwicklung einer neuen Ingenieurbiologie? Dornburg (Forum W).

Contact: Berlin-Brandenburgische Akademie der Wissenschaften, Interdisziplinäre Arbeitsgruppe Gentechnologiebericht, Jägerstr. 22/23, 10117 Berlin.

Too Early or Too Late? The Assessment of Emerging Technology

Michael Decker

Abstract

Which topics is technology assessment concerned with? And when? In response to this question, David Collingridge has pointed out a dilemma between using such research at relatively early or later points in time: In the first case, one does not have enough knowledge although the scope for influencing events is still large, while in the second case knowledge for evaluating an issue is available, yet the process of technical development is already so advanced that the space for options to influence events is distinctly limited. This dilemma continues to exist but technology assessment (TA) has responded methodologically to the different challenges and in particular has developed a tool set for assessing what is called the new and emerging technologies (NEST). In this paper I will present criteria with which we can answer the question whether TA should concern itself with an emerging technology. Finally, I will derive some consequences from these criteria as to if and how TA should deal with epigenetics.

M. Decker (✉)
Institute for Technology Assessment and Systems Analysis (ITAS), KIT—The Research University in the Helmholtz Association, Karlstraße 11, 76133 Karlsruhe, Germany
e-mail: michael.decker@kit.edu

© Springer Fachmedien Wiesbaden GmbH 2017
R. Heil et al. (eds.), *Epigenetics*, Technikzukünfte, Wissenschaft und Gesellschaft / Futures of Technology, Science and Society,
DOI 10.1007/978-3-658-14460-9_3

31

1   Introduction

The term "emerging" as in emerging technology refers—as can be found in
Merriam–Webster's Unabridged Dictionary—"to rise from," to "become manifest," and "to come forth."[1] These words conceptually designate a point in time at
which this occurs. Yet if we, as would seem natural, proceed from the assumption
that new technologies are constantly being developed and if we add the word
"new" to "emerging" as in the expression "new and emerging (sciences and)
technologies" (NEST), as is used in particular in the language of EU research
policy, then the question arises as to who is observing this ongoing process of
technology development and for which reason in order to designate those that are
relevant to him as newly emerging. The relevance also includes in particular the
innovativeness of a technological innovation. While, for example, an engineer
working on the design of cylinder heads for a gasoline engine views each modified
form of a cylinder head as a new technology that is relevant to him, for another
engineer who works on automobile drive systems in general, such relevant novelty
will only come with a new form of technology such as a rotary engine. Different
criteria for assessing relevance obviously play a special role, and these criteria still
have to be developed against the background of the purposes associated with this
observation.

TA is concerned—literally—with assessing the consequences of a type of
technology. In addition to the technical aspects, it also takes precisely the nontechnical aspects of technology into account. At issue are thus the positive and
negative consequences and, in particular, the (un)intended, (un)desired, and the
primary and secondary consequences of technology (Gloede 2007; Decker 2013).
These consequences are first taken into account and then put in relationship to one
another conceptionally, in order to develop alternatives for acting. Given the fact
mentioned above that new types of technology are constantly reaching different
levels of innovation, TA as an institution has to face the problem of selecting its
objects of study, if only for the trivial reason that it does not possess infinite
financial resources for conducting TA projects. It therefore must have a selection
procedure at its disposal that permits it to find the most relevant types of technology from a TA perspective. For these decisions about relevance to be made in a
transparent manner, the corresponding criteria have to be explicated (Decker and
Fleischer 2010).

---

[1]Similar meanings—"*auftauchen*," "*sich hervortun*," and "*emporkommen*"—can be found
for the word "*emergierend*" in *Duden*, a German dictionary.

This paper will present criteria on the basis of which these decisions about relevance to TA can be made. They were determined in the course of the project ITA Monitoring: Identification of New Topics for the Analysis of Innovation and Technology,[2] which was financed by the German Federal Ministry of Education and Research (BMBF).[3]

2   Criteria of Relevance for Technology Assessment

Following the fundamental idea of TA that was outlined above, it is possible to formulate in very general terms that TA takes on a technological topic when there is the justifiable expectation that a specific type of technology is linked to social, political, and economic consequences (Bröchler and Simonis 1998; Gethmann and Sander 1999). Initially, this is a weak means of making a distinction since starting at a certain level of application every type of technology has social and also economic consequences. It is therefore necessary to elaborate in greater detail the criteria of relevance for TA to occupy itself with a specific type of technology. The criteria described in the following were developed in the framework of the research project ITA Monitoring mentioned above. They can, however, be considered typical for technology assessment that views itself as being problem oriented (Bechmann and Frederichs 1996; Grunwald 2002) and for this reason must initially prepare a systematic description of the existing problematic social situation. Such a description can be oriented on these criteria. To this extent, these criteria can be employed analogous to, for example, the *dimensions* developed in the TAMI EU project (Bütschi et al. 2004, pp. 20ff.).

The ITA dimensions are (Decker et al. 2012, pp. 9ff.):

1. Technical-scientific dimension
   A TA action is generally always focused on a specific technology that is to be assessed. The analysis depends on this scientific-technological development and is supposed to present the utility of the technology and its potential for solving a problem, just as it is the state of development, the scientific knowledge involved, and the potential consequences.

---

[2]Innovation and technology analysis was introduced by the BMBF in 2001 (BMBF 2001).

[3]The thoughts presented here were developed out of the two reports of the project (Decker et al. 2012, 2014), in which additional references can be found.

2. Economic dimension

An analysis of the current or future economic significance of the identified technology at the national and international level is also of importance (market potential, savings potential, impact on the labor market, etc.). In the context of such an analysis, the value chain and the utilization chain should also be evaluated, for example being examined as to whether there are any indications of possible obstacles to utilization (procedures, products, patents, etc.) or market drivers in connection with the identified technology. Another important criterion is the consideration of competition and market regulations.

3. Ecological dimension

Another characteristic aspect is the assessment of the ecological dimension, which concentrates on approaches to solutions that are relevant for the environment, such as the environmental compatibility of products and manufacturing processes and the recyclability of products. In doing so, the focus should also be on the possible potential for conflict, such as the statements by environmental groups for and against the identified technology.

4. Social dimension

The social perspective of a technical development and a field of technology is another decisive factor in the analysis of innovation and technology. First, there is the question of the degree to which the identified technology will contribute to an increased quality of life or prosperity, the degree to which demand for it will manifest itself with regard to quality of life and prosperity, and whether the public will recognize this. Since social approval or rejection also plays an important role in the market penetration of a specific technology, it should also be studied whether and who, in the context of the technology or expressions of need, discusses social consequences and whether the potential for conflict or consensus is just as prominent as are socially relevant uncertainties/risks and opportunities. It should therefore be examined in the framework of an ITA whether the technological development affects patterns of behavior that are culturally engendered and traditional in nature (such as education, media usage, and mobility).

5. Political dimension

The assessment of the technological solution also depends on the concrete political situation in which the technological development takes place, since problematic situations addressed by ITA are generally politically relevant. The nature of the relevance depends on the stage of the political decision-making process and on the nature of the ongoing political discourse. If a specific technology or a need has already been recognized at the political level, the task is to identify the political actors participating in the political discussion. Of most importance here are the constellations or alliances of actors and whether the topic is being discussed in the context of conflict or consensus. The need for regulation, legislation, and support should also be analyzed in this connection. It should furthermore be considered whether the topic is largely national in character or whether international connections play a role. The influence potentially exerted at the political level (i.e., the state or federal governments, the EU, or the UN) must be examined just as the topic's connection to the geopolitical situation is, particularly if the latter could affect Germany's fundamental situation.

6. Health dimension
   Not least important, any indications of approaches to (or promises of) solutions with regard to health factors should be taken into consideration. This also includes the analysis of the potential consequences for health and the potential dangers (e.g., which indications are there of possible human toxicity in connection with the identified technology?) and of any positive or negative comments about the technology by health and patient organizations.

The need to be able to operationalize these dimensions in a real monitoring process stood in the forefront during their formulation. This means that the dimensions should be formulated so that they can be easily imparted to those individuals who conduct the monitoring process and that they make it possible to present the selection process in the methodological procedure in a transparent manner. In this operationalization, it was less relevant, for example, that the dimensions are not disjoint. Overlapping results do not pose a problem since it is not important which dimension a certain topic is found by. A topic could, thus, be raised by a patient organization and be found both by the dimensions "health" and "society". It was important, in contrast, that the relevant topics were found at all, which ultimately speaks in favor of the dimensions producing topical overlap to ensure that the criteria do not leave any topics undiscovered.

It is also evident that criteria are mentioned in the dimensions that can be directly attributed to specific scientific disciplines. This is most obvious with regard to the economic dimension. Other dimensions are indirectly linked to corresponding scientific disciplines without being subsumed completely by their disciplinary perspective, such as the health dimension is not subsumed entirely by medicine. If one understands the dimensions given above as the point of reference for describing a concrete problematic situation and decides to prepare suggestions for solving the problem on a scientific basis, a decisive methodological step is then to transform the problem into a question that can be worked through scientifically (Defila and Di Giulio 1999; Decker 2007). In doing this, a multidisciplinary process can be assumed to be the normal situation, since a social problem generally cannot be attributed monocausally to a scientific discipline when searching for a solution. The solutions have to be developed in interdisciplinary and transdisciplinary collaboration, during which the transdisciplinary research furthermore puts the emphasis on the capacity to implement the generated knowledge and on its robustness, which can be achieved by involving further actors, i.e., extended peers (Nowotny et al. 2001; Gibbons et al. 1994; Funtowicz and Ravetz 1993, 2001; Ravetz and Funtowicz 1999). Decisive in these new forms of generating knowledge is that the problem-solving potential can actually be reached in the real world. In the above-mentioned sense, this can be understood as a kind of reverse transformation from science into the real world. The degree to which the results of the transdisciplinary research contribute to solving the problem then becomes a criterion of quality, if not the decisive one.

From the perspective of transdisciplinary research on the impact of technology, two aspects are thus of particular importance, namely the identification of the scientific disciplines relevant for determining the solution to the problem and also of the relevant nonscientific actors.

3 Discussion

In the Introduction, a temporal component was identified for the concept "emerging technology," namely the point in time at which a specific technology gets caught in the grid of criteria and becomes conspicuous. This can also be interpreted as the earliest possible point in time at which research into the impact of a specific technology is possible since prior to this it could not have been perceived.[4] Collingridge (1982), however, has pointed out in this connection that this may be too early to concern oneself with the consequences of a new and emerging form of technology. According to Collingridge, TA research finds itself in a dilemma since it has so little knowledge about this technology at its disposal at this early stage of its development that it cannot make any reliable statements about its possible impact. At a later stage of the technology's development, this knowledge may be available or could be created, and from this knowledge about the impact of the technology it would be possible for corresponding options for taking action to be deduced. Yet at this later time, the process of technical development is already so advanced that our ability to modify the technology would be much more limited than at an earlier stage of development. Ultimately, according to Collingridge, the decisive turning points would already have been set and larger changes in direction could only be achieved at relatively high costs.

The polar contrast presented in the Collingridge dilemma constitutes a true methodological challenge to research on the impact of technology. Yet it has not led to the termination of such research for being conducted either too early and thus being unable to develop any options or too late and then being unable to identify any alternatives that can be achieved. On the contrary, it has led to an expansion of TA's methodological spectrum and even to the level of development of the specific technology being taken into consideration in the description of the problem (see the innovation dimension in the TAMI criteria). The methodological procedure developed out of both the problem description and the concrete aims of the research into the impact of the technology then has to do justice to the—in this case early—level of innovation by, for example, identifying the relevant gaps in our knowledge and registering the need for research or developing further specific monitoring criteria for appropriately analyzing a process of technological development. In this sense, the short studies of the ITA monitoring project,[5] after describing the problem with reference to the above-mentioned ITA dimensions, also suggested concrete methodological procedures appropriate to the situation, i.e., indicating how the individual issues derived from the problem description

---

[4]In the discussion of how early a new form of technology can be perceived, weak signals have been mentioned over and over again, i.e., the first, very faint signals that would make a very early form of perception possible if one had a corresponding detection system. See Decker et al. (2014).

[5]They were prepared on the topics of electromobility, intelligent power grids, sticky information, nonmedical uses of the neurosciences, fresh food in the trash (the video of the same name in German—*Frisch auf den Müll*—was titled Taste the Waste in English), service robots in care taking, and the social prerequisites of efforts to achieve technical enhancements of human capacities.

could be approached methodologically. In the process, it becomes apparent that—as is presumably the case in every scientifically framed problem—several possible methodological procedures can be justified as being capable of identifying solutions, even partial ones.

Against this background, I would like to take a look at the conference week held on the topic of epigenetics. The arguments based on considerations of epigenetics make it very plausible that we need a better understanding of epigenetic connections in order to develop solutions to very pressing societal problems, such as those named in the call for the submission of abstracts to this retreat: "The susceptibility to diseases and precisely to widespread ones (such as cancer and obesity) and to mental disturbances and the sensitivity to environmental toxins can be traced back in part to epigenetic effects." At the same time, it appears to be possible to modify the nature and degree of these epigenetic effects by means of changes in behavior. This leads me to another point that is relevant for the TA of new and emerging technology: the framing of the concept in the sense that one can decide what belongs to it—and thus becomes an object of TA—and what does not. This can—but does not need to be—a definition of a specific technology, and for the use of this definition to be justified as being appropriate for the purposes of assessing the technology. This means that the "definition" leads to the desired distinctions that distinguish epigenetics from other, possibly similar subjects. In some circumstances, it can even suffice to explicate and ground the intended differentiation. This takes the fact into account that TA approaches a new form of technology from the outside. This aspect was, for example, given a relatively large amount of attention during the early stages of nanotechnology and of the TA of nanotechnology. It was also apparent in this discussion that it is generally not the protagonists of a specific technology (e.g., the nanotechnologists) who—looking out from their inside perspective, as it were—need a more unambiguously defined demarcation. Such a demarcation is however important if you want to regulate nanotechnology or formulate targets for nanotechnology on the basis of research policy. In such a case, it must be possible to operationalize the demarcation so that it can be employed as a basis for making concrete decisions. I can easily imagine that such demarcations and their justifications will still have to be achieved for epigenetics as a field of research.

In this sense, a problem-oriented and thus interdisciplinary and transdisciplinary treatment of the topic of epigenetics is desirable socially and sensible in terms of health policy. Furthermore, the procedure selected here is expedient for viewing epigenetic effects from the perspectives of different scientific disciplines. Whether and to what degree research on the impact of technology plays a central role in considerations of further work on epigenetics are uncertain and—let me recall the technical dimensions of the criteria grid—depend on which roles can be attributed to technical artifacts or technical processes in this connection.

References

Bechmann, G., & Frederichs, G. (1996). Problemorientierte Forschung: Zwischen Politik und Wissenschaft. In: G. Bechmann (Ed.), *Praxisfelder der Technikfolgenforschung. Konzepte, Methoden, Optionen* (pp. 11–37). Frankfurt a.m.: Campus.

BMBF (Bundesministerium für Bildung und Forschung BMBF, Ed.) (2001). Innovations- und Technikanalyse. Zukunftschancen erkennen und realisieren. Bonn.

Bröchler, S., & Simonis, G. (1998). Konturen des Konzepts einer innovationsorientierten Technikfolgenabschätzung und Technikgestaltung. *TA-Datenbank-Nachrichten, 7*(1), 31–40.

Bütschi, D., Carius, R., Decker, M., Gram, S., Grunwald, A., Machleidt, P., et al. (2004). The practice of TA. Science, interaction and communication. In: M. Decker, & M. Ladikas (Eds.), *Bridges between science, society and policy. Technology assessment-methods and impact* (pp. 13–55). Berlin: Springer.

Collingridge, D. (1982). *The social control of technology.* New York: St. Martin's Press/London: Pinter.

Decker, M. (2007). *Angewandte interdisziplinäre Forschung in der Technikfolgenab- schätzung.* Bad Neuenahr, Ahrweiler: Europäische Akademie (Graue Reihe 41).

Decker, M. (2013). Robotik. In: A. Grunwald (Ed.), *Handbuch Technikethik* (pp. 354–358). Stuttgart/Weimar: Metzler.

Decker, M., & Fleischer, T. (2010). When should there be which kind of technology assessment? A plea for a strictly problem-oriented approach from the very outset. *Poiesis & Praxis, 7,* 117–133.

Decker, M., Fleischer, T., Schippl, J., & Weinberger, N. (Eds.). (2012). *Zukünftige Themen der Innovations- und Technikanalyse* (p. 7605). Methodik und ausgewählte Ergebnisse. Karlsruhe: KIT Scientific Publishing (KIT Scientific Reports.

Decker, M., Fleischer, T., Schippl, J., & Weinberger, N. (Eds.). (2014). *Zukünftige Themen der Innovations- und Technikanalyse* (p. 7668). Lessons learned und ausgewählte Ergebnisse. Karlsruhe: KIT Scientific Publishing (KIT Scientific Reports.

Defila, R., & Di Giulio, A. (1999). Evaluationskriterien für inter- und transdisziplinäre Forschung. Projektbericht. In Schwerpunktprogramm Umwelt Schweiz (Ed.), *Trans- disziplinarität evaluieren – aber wie? Panorama Sondernummer 99* (pp. 3–15). Bern: Interfakultäre Koordinationsstelle für Allgemeine Ökologie.

Funtowicz, S., & Ravetz, J. R. (1993). Science for the post-normal age. *Futures, 25,* 739–755.

Funtowicz, S., & Ravetz, J. R. (2001). Post-normal science. Science and governance and conditions of complexity. In M. Decker (Ed.), *Interdisciplinarity in technology assessment: Implementation and its chances and limits* (pp. 15–24). Berlin/Heidelberg: Springer.

Gethmann, C. F., & Sander, T. (1999). Rechtfertigungsdiskurse. In A. Grunwald & S. Saupe (Eds.), *Ethik in der Technikgestaltung* (pp. 117–151). Berlin/Heidelberg: Springer.

Gloede, F. (2007). Unfolgsame Folgen. Begründungen und Implikationen der Fokussierung auf Nebenfolgen bei TA. *Technikfolgenabschätzung – Theorie und Praxis, 16*(1), 45–54.

Gibbons, M., Limoges, C., Nowotny, H., Schwartzman, S., Scott, P., & Trow, M. (1994). *The new production of knowledge: Dynamics of science and research in contemporary societies.* London: SAGE.

Grunwald, A. (2002). *Technikfolgenabschätzung – Eine Einführung.* Berlin: edition sigma (Reihe: Gesellschaft – Technik – Umwelt. Neue Folge 1).

Nowotny, H., Scott, P., & Gibbons, M. (2001). *Re-thinking science. Knowledge and the public in an age of uncertainty.* Cambridge: Polity.

Ravetz, J. R., & Funtowicz, S. (1999). Post-normal-science—an insight now maturing. *Futures, 31,* 641–646.

Author Biography

Michael Decker is Full-Professor of Technology Assessment at the Institute for Philosophy and Head of the Institute for Technology Assessment and Systems Analysis (ITAS). Interdisciplinary Research in Technology Assessment (TA) (Decker, M. (ed.) Interdisciplinarity in Technology Assessment. Implementation and its Chances and Limits. Berlin Heidelberg Springer 2001) Methodology of TA with a special focus on the impact of TA in advising societal actors (Decker, M.; Ladikas, M. (eds.) Bridges between science, society and policy. Technology Assessment—Methods and impacts. Berlin, Heidelberg, New York: Springer 2004) and TA of ICT and robotic systems (Decker, M., Gutmann, M. (eds.) Robo- and Informationethics. Some fundamentals. LIT-Verlag Wien, 2012) are his areas of expertise and are still indicating his research interests such as theory and practice of Technology Assessment; research on the societal consequences of nanotechnology and robotics; and concepts of inter- and transdisciplinary research.

Contact: KIT—The Research University in the Helmholtz Association, Institute for Technology Assessment and Systems Analysis (ITAS), Karlstraße 11, D-76133 Karlsruhe.

Epigenetics and Genetic Determinism (in Popular Science)

Sebastian Schuol

Abstract

When the regulative influence of the environment on genes is seen in the postgenomic discourse as evidence against genetic determinism, epigenetics seems to solve the problem. This interpretation is premature. The argument of gene–environment interaction refutes only a simple version of genetic determinism, whereas a more complex version of it not only persists, but is actually promoted by the mask of the "solution". The reason for this covert genetic determinism is an asymmetric perception of gene–environment interaction. Often popular scientific representations employ different information criteria. While genetic information is there understood in the sense of an intentional instruction, epigenetic information is thought to apply only to its regulation. For a comprehensive refutation of genetic determinism, reference to the interactions between genes and environment is insufficient and therefore the whole process of the development of information must be taken into account.

1   Introduction

Considering the positive headlines in popular science articles, critical discussion of epigenetics and genetic determinism (GD) seems misplaced. The German magazine *Der Spiegel* (2010, p. 1) proclaimed for example the final "victory over genes"

S. Schuol (✉)
National Center for Tumor Diseases (NCT) Heidelberg, Im Neuenheimer Feld 460, 69120 Heidelberg, Germany
e-mail: Sebastian.Schuol@med.uni-heidelberg.de

© Springer Fachmedien Wiesbaden GmbH 2017   41
R. Heil et al. (eds.), *Epigenetics*, Technikzukünfte, Wissenschaft und Gesellschaft / Futures of Technology, Science and Society,
DOI 10.1007/978-3-658-14460-9_4

because of the gene regulatory impact of the environment. But there is room for doubt. In the following a series of conceptual differentiations outline the relationship between epigenetics and GD, and a proposal for refutation of the latter is made.

In the first section a distinction is made between the earlier and the modern concepts of epigenetics. While epigenetics initially was introduced as an alternative to a deterministic view of genetics, this aspect is currently losing its significance. In molecular genetics, epigenetics has changed its meaning, referring to a spatial relationship that regulates gene activity. Insofar as this model is actually the dominant one, the following analysis is limited to the molecular concept. In the second section, a differentiation between two kinds of GD shows that the "epigenetic" refutation of GD refers only to a simple version of the latter. For a better understanding of the second version and its ethical impact, the concept of information is important, and therefore different modes of representing genetic and epigenetic information will be shown. In the last step, a proposal for the refutation of GD is made independent of epigenetics. Since GD is based on a theory-driven perception of information, every empirical refutation must fail. In this article a solution can only be indicated.

2   Epigenetics

There are at least two different concepts of epigenetics. While the embryologist and geneticist Conrad Waddington introduced the term "epigenetics" to explain development in a non-gene deterministic manner, the current molecular concept of epigenetics is increasingly losing sight of this aspect.

2.1   Waddington's Epigenetics

Historically the beginning of epigenetics is located in the era of the *classical genetics*, which had no material concept of the gene and its regulation. In contrast to our current understanding, the former concept of epigenetics refers neither to DNA nor to molecular control mechanisms. At that time, genetic research was carried out at the level of the phenotype. Waddington's research is devoted to development (experimental embryology), which has a procedural view of genetics. While genetics explored the transgenerational transmission of phenotypical characteristics and therefore put the intermediate stages of development in a black box, Waddington's epigenetics is devoted to this "in between". In retrospect, he writes: "Some years ago (e.g., 1947) I introduced the word 'epigenetics', derived from the

Aristotelian word 'epigenesis', which had more or less passed into disuse, *as a suitable name for the branch of biology which studies the causal interactions between genes and their products which bring the phenotype into being*" (Waddington 1968, pp. 9f., italics add).

This sentence is often cited in the current literature on epigenetics, placing the focus on the second, highlighted half of the sentence because it is suitable as a definition of Waddington's epigenetics. This emphasizes the connectivity of the original to molecular epigenetics, which indeed turns to similar questions of research but loses sight of the fact that Waddington understands his research in a specific tradition. He introduced the concept epigenetics as a position against GD.

Waddington refers to the theory of *epigenesis*, which is a counter-theory to *preformation*, a forerunner of GD. Epigenesis and preformation each claim to explain the development of individuals but in doing so they differ drastically from each other. While *epigenesis* presupposes an initial shapeless mass from which the organism gradually evolves along environmental cues, preformation presupposes a preformed substance so that the environment has the mere status of a dietary substrate. Waddington referred to this contrast when facing preformistic genetics, presenting his "epigenetic" understanding of development and combining both approaches in the name of the synthesis project "epi-genetics". He knows about the importance of the genes. Development is not only caused by environmental cues, but runs in a genetic framework. For him, this poses a serious problem of compatibility. How is such a dynamic process as development possible despite the constancy of the genetic frame? To solve it he develops a new concept of genetics. In *The Cybernetics of Development* Waddington (1957, p. 29) provides a topographic metaphor to explain development. This *epigenetic landscape* has two parts.

In the upper part, the formation of the epigenetic landscape is mechanically explained by a lower perspective. Genes, or more precisely the products of genes, interact and bring forth a netlike structure. The shape of the epigenetic landscape is not static but generated in a fluid manner by (1) the state of the organismic development, (2) the genetic disposition, and (3) the environmental stimuli. According this logic, developmental information does not exist on any of these levels, but arises as a product.

In the part above, the epigenetic landscape is shown in a view from above, and a bullet marks the developmental state. The texture of the landscape determines the course of the bullet. Sensitive developmental phases are shown as bifurcations. Although gravity is leading the bullet in the right path, the form of the left one is losing its developmental power. Following this metaphor, there are an infinite number of paths. The course of development is determined: Waddington understands epigenetics as developmental cybernetics, in which all the factors influence each other in a feedback process. In this epigenetic landscape, the ball rolls to its final state, i.e., the differentiated tissue.

On this basis, Waddington's epigenetics cannot be understood as monocausal determination. Whilst explaining development as a process of emergence epigenetics is going even beyond gene–environment dualism; Waddington (1952) was the first who thought in terms of developmental systems. He chose the term epigenetics to mark his turning away from the gene deterministic concept of genetics.

2.2 Molecular Epigenetics

The subsequent conceptual changes do not follow a linear development. Due to the lack of applicability in the following era of molecular genetics, Waddington terminated his project. His epigenetics was conceptualized in the paradigm of classical genetics and its focus on phenotypes, but it was subsequently reinterpreted in a chemo-physical manner. While Waddington's epigenetics represents a theory of genetics (the concept of the gene depends on that of epigenetics), the discovery of the materiality of inheritance has changed this into its opposite (van Speybroeck 2002, pp. 78f.). Now the concept of epigenetics depends on that of the molecular gene and its localization on DNA. Recent history shows that epigenetics first included DNA and then excluded it, in this way reaching that specific relationship that influences our current spatial concept.

Since Waddington's epigenetics unfolded no lasting scientific influence subsequent research lacked a common focus. It has only been in the last 25 years that a unitary concept of epigenetics has arisen. In this context, the famous work on cellular memory by the molecular geneticist Robin Holiday counts as a key stimulus (Jablonka and Lamb 2002, p. 87). Looking for a mechanism of cell differentiation, Holiday defines epigenetics as "the study of the mechanisms of temporal and spatial control of gene activity during the development of complex organisms" (Holliday 1990, pp. 431f.). Since the DNA in all the cells of an organism is the same, it cannot contain the information for cell differentiation. Holliday suspected the "memory" beyond genetic information. But since development includes differentiation processes based on modifications of DNA (e.g., the immune system), his epigenetics concept includes DNA. This concept was changed after these modifications were excluded as a cause of cellular memory. To integrate this into research, Riggs et al. (1996) defines epigenetics as "the study of mitotically and/or meiotically heritable changes in gene function that cannot be explained by changes in DNA sequence" (ibid., p. 1). While DNA was being definitely decoupled from epigenetics (modifications of the base sequence), a further conceptual change was taking place. Epigenetics' field of reference expanded beyond the realm of development to that of inheritance. Since DNA was excluded as an information carrier, the question arises as to the cause of transgenerational effects.

The desired information carrier is found in a spatial relationship to the DNA. All epigenetic processes are based on one principle, which is determined by two mechanisms. The general mechanism relates to condensation of DNA. In its functional state DNA is organized together with proteins, and both form a functional complex. This nucleoprotein is called chromatin and exists in two states. While *heterochromatin* refers to dense DNA packing, this is loosened in *euchromatin*. The two states have a regulatory effect. Due to the compression of DNA, the accessibility of transcription factors is regulated at specific genomic loci, which favors (*euchromatin*) or disables (*heterochromatin*) expression. This general mechanism is controlled by further mechanisms on the level of DNA (methylation), RNA (RNA interference), and proteins (histone acetylation) (Youngson and Whitelaw 2008).

The methylation of the DNA base cytosine illustrates perfectly the spatial concept of molecular epigenetics. The modification of the cytosine plays a crucial role in DNA condensation, but can also regulate gene activity independent of this. At the front of most genes is a high incidence of the base sequence cytosine–guanosine. The cytosine in these CpG islands (cytosine-phosphatidyl-guanosine) can be modified chemically by methylation. This modification is reversible and does not change the base sequence of the DNA, but has a regulative function. Spreading from the DNA strand this methylation blocks the binding of transcription factors to the promoters of genes and thereby prevents their expression. Only unmethylated genes are translated into proteins.

Modern epigenetics focuses on molecular factors like methylation, which lie above the DNA strand and regulate gene expression. Consequently, the prefix "epi" does not refer to the ancient concept of epigenesis but only to the Greek word that means on/over/above. This spatial understanding reduces the earlier concept, and Waddington's interest in merging epigenesis and genetics is lost from sight. While his synthesis refers to development of the phenotype as an emergent process and is directed against GD, epigenetics in our age of molecular genetics may indeed still be interested in development, but it now refers only to a spatial relationship exerting a regulative function.

3   Genetic Determinism

GD differs from *methodological determinism*, which presupposes a causally closed world and in turn facilitates scientific research. GD is based on the assumption that the phenotype is genetically determined. Two versions of GD and its relation to epigenetics are discussed in the following sections.

3.1   Simple Genetic Determinism

The *simplest* version of GD (sGD) is sometimes set equal to the *central dogma of molecular biology* (Tappeser and Hoffmann 2006). This dogma describes a unilinear flow of information: the base sequence of a gene is transcribed from DNA to messenger RNA which then is translated in the peptide sequence of a protein (Crick in 1958, p. 153). This relationship is unilinear since the base sequence of DNA cannot be modified by the somatic part (protein, RNA). sGD assumes also that only genes characterize the phenotype, but it goes beyond the central dogma in two respects. The first is in relation to the objective. The determination does not end at the molecular level with the peptide sequence of proteins, but extends to the phenotype. The second is with respect to the starting point. The central dogma's statement of directionality is expanded in a statement of causality. The sGD makes the additional statement that the information flow is genetically initiated. Analogous to preformationism, in the context of the sGD the environment only has the status of a food substrate; it exerts no influence on the expression of the phenotype. While it is important that there is an environment, it does not matter what exactly it is. Consequently, the expression of the phenotype in differing environments remains uniform; only the genotype determines the phenotype. For some, sGD is confirmed by genetic diseases with high penetration, such as autosomal dominant inheritable Huntington's disease. But the universal validity of sGD cannot be derived from the existence of this rare genetic disorder (Bartram et al. 2000, pp. 11f.). However, results of epigenetic research refer to the crucial role of environmental conditions in development and thus erode the key message of sGD. For this research, the genome does not have the status of an information monopoly, and a dialogue between genes and environment is the cause of development. Which genes are expressed when and to what extent is mainly controlled by the epigenetically active intracellular milieu, which can be seen as a mirror of the extraorganismal environment (Cooney 2007). In the context of gene regulation, a flow of information thus takes place from the level of the environment to the gene.

That the environmental influence on gene regulation also exerts significant influence on the expression of phenotypes has been shown in numerous studies with plants and animals, but also with humans. Here, the epigenetic priming of physiological control systems (neuro, hormone system) during early developmental stages is of importance, since it has an impact on rampant lifestyle diseases such as metabolic and cardiovascular disorders (Godfrey et al. 2007).

Epigenetic research shows that environmental influences contribute actively to the development of the phenotype. This means that the sGD should be dropped and that there is no unilateral determination. The consequent conclusion of a bilateral flow of information forces us to distinguish genetic from epigenetic information. That epigenetic regulations cause only spatial modifications of the DNA, without changing the underlying genetic information of the base sequence, has been shown. Although the backflow of epigenetic information refutes the thesis of sGD, the *central dogma* remains valid.

3.2   Covert Genetic Determinism

To my knowledge, the version of GD that I now refer to as *covert* GD (cGD) has not yet been addressed in the literature. It might be helpful to introduce it in contrast to sGD. The best example for the refutation of the sGD comes from twin research. Since its beginning in the nineteenth century (Galton 1875), the nature–culture debate has played a central role in this research paradigm and is taken up by epigenetic research. Research with monozygotic twins is particularly interesting, since any phenotypic variations can only explained by environmental effects because the twins share the same genome. Fraga et al. (2005) were the first who showed that the gene activity of monozygotic twins is similar at the age of 3 years but then diverges in direct relation to age. They assume that the reason is deviation from the early common environment. Accordingly, etiological studies investigate the epigenetic differences between twins on the basis of their different disease profiles. In this paradigm the cause of the disease has shifted to the environment, which—according to their assumption—is causing a pathogenic deregulation of the gene activity.

Thus, several phenotypes can be developed from one genome depending on the environment, but does this refute GD? In heredity research these interactions are mostly investigated quantitatively, and the relative proportion of the genetic or environmental effect is calculated. But a qualitative aspect is of interest here, which is the issue of how genetic or epigenetic information is discussed in popular science. In this interactional talk it remains unclear what kind of information is meant. The philosophers Mahner and Bunge (2000, pp. 275ff.) show that information in biological discourse can have six different meanings. The word is ambiguous. In the following, the distinction between *causal* and *intentional* (Sterelny and Griffith 1999, p. 101; Griffith 2006, pp. 182ff.) or *semantic* information (Griffith and Stotz 2013, pp. 160ff.) is important.

Causal information refers to the degree of order (negentropy) of a system, and this can be described mathematically. Its informational value is measured in bits. If, as is the case for data processing systems that operate on the basis of binary codes, only one of just two states (power on or off) is possible, this corresponds to the informational value of 1 bit. In contrast, *intentional* or *semantic* information reflects our everyday concept of information, where information means instruction. Since only this form of information has the meaning of a blueprint, GD refers to it (gene x for feature X).

When talking of gene–environment interactions, we presuppose that both parts are of equivalent importance (Griffith and Knight 1998, p. 254). But such parity is rarely guaranteed. Different degrees of importance are usually associated with the respective parts. Since the methylation of the cytosine determines the state of genetic activity, it is considered regulatory information. In this view, at the epigenetic level each gene is binary coded and has informational value of 1 bit (gene on–off). Epigenetic information carriers like methylation are accordingly associated with *causal information*. In contrast, the base sequence of DNA is treated as morphogenetic information. Although genetic information is itself controlled, only the base sequence of genes "contains" information concerning genesis and is therefore associated with *intentional information*. Understanding gene–environment interactions in this sense serves cGD. Since gene activity is determined in dependence on the environment, the genome does not determine the phenotype in the ratio 1:1 (sGD) but 1:X. Here the phenotype is determined by the gene set of the genome that has been epigenetically activated.

I would like to warn against generalizing. Neither twin research nor epigenetics per se should be accused of genetic determinism. Here I have referred only to the problematic manner characteristic of popular science articles in which epigenetic and genetic information has been presented. An example might help to make this clearer. Popular science often facilitates its explanation of scientific concepts by using everyday images, e.g., the concept of a "library" is often used as a metaphor to explain epigenetics (Fischer 2013; Vivamus 2014; Staege 2014). A library accordingly represents the genome of an organism, and its books the genes. Only specific books contain information about a specific issue, e.g., model airplanes. In the metaphor, the selection process stands for epigenetic regulation, and loaned books for activated genes; they contain the required information. Analogous to books left on a shelf, the information of the inactive genes has no effect on the phenotype. Even though both the selection process (gene regulation) and the books (genes) are important and no part may be omitted, only the borrowed books contain the instructions actually employed to build airplanes. According to this popular metaphor for epigenetics, only books (or their pendant, genes) contain *intentional information.*

When popular science articles proclaim that epigenetics has refuted GD and therefore use metaphors that represent genes as intentional information, such refutations promote cGD. This has further consequences. Since this refutation of sGD pretends to have solved all the problems, such a "solution" covers GD, opening the door for it to determine our thinking about genes subliminally. As we have seen, the emphasis on gene–environmental interactions refutes sGD, but not cGD.

3.3   From Fatalism to Activism

What is the problem concerning cGD? Since the ethical core problem of GD is located at the level of freedom gene fatalism is problematic (Midgley 1984, p. 111). Whoever believes that health is genetically determined has no reason to change their risky health habits. It is important to see that the objective nonexistence of GD is irrelevant here. The well-known Thomas theorem from sociology states that if people define situations as real, they become real in their consequences (Thomas and Thomas 1928, p. 572). The mere belief in GD favors unhealthy actions, and the consequences of these actions reinforce this belief in the sense of a self-fulfilling prophecy.

Because of its emphasis on gene–environment interaction, cGD does not promote such fatalism, and for this reason it does not seem a problematic case of applied ethics but just a theoretical subtlety. The popular science headlines suggest that epigenetics solves all the problems of GD, and indeed since epigenetics is understood in the sense of a liberation, its reception is mostly positive. It is helpful to consider this development in more detail. According to the new epigenetic discourse on lifestyle, it is considered that an individual can direct its gene activity in a health-promoting way by choosing the "right" environment (Seitz and Schuol). The former gene fatalism is turning into a gene activism. What first appears to be environmental determinism is on closer examination cGD since the gene activity desired by an individual determines the individual's choice of a particular environment. Prompts that fall in this context such as "to be master of his own destiny" (Markert 2008, p. 100) show that the cause of illness shifts from a level of an inaccessible nature to the volitional level of action. The scientific status of our knowledge and any doubts as to whether gene activity can ever be "directed" do not matter. The pure belief in this power and the following corresponding actions change social thought. According to this epistemic shift, diseases are changing into a problem of self-responsibility. I have shown elsewhere that this view leads to problems of social injustice (Schuol 2014). While it seems plausible to trace the refutation of gene fatalism back to the research findings of epigenetics, this does not solve all the problems of GD. They are simply moved. There is an association between the postgenomic "mutation of the gene discourse" (Lemke 2002) and a mutation of the formerly problematic areas. The crucial point of this change remains—at least in public presentations—the exaggeration of the gene as embodying intentional information that is special and stands hierarchically above gene regulation. For these reasons sGD will not play such a big role in future ethical discourse, in contrast to cGD. Since cGD is caused by a very specific understanding of information, its resolution cannot start at the empirical level of epigenetics but must start at the theoretical level of the concept of information.

4   Against Genetic Determinism

While Kitcher (2001) sees the origin of GD in a problematic oversimplification of scientific knowledge and holds it principally for preventable, Oyama (1985) sees its origins as being systematic. According to Oyama's view, Kitcher's demand that biological themes should be presented in their complexity will not help insofar as conceptual thinking is concerned (Griffith 2006). Without careful reconceptualization, any refutation would be doomed to failure, and GD would soon revive because of its conceptual roots. Oyama's proposal for a final solution to the problem posed by GD is the following:

> What we need here, to switch metaphors in midstream, is the stake-in-the-heart move, and the heart is the notion that some influences are more equal than others, that form, or its modern agent, information, exists before the interactions in which it appears and must be transmitted to the organism either through the genes or by the environment (Oyama 1985, pp. 26f.).

Crucial in Oyama's proposal is the procedural understanding of gene–environment effects, where information is seen as product of interaction. Before interaction takes place, neither the environment nor genes contain information. While Oyama's solution seems easy on an epistemic level—it just requires rethinking information from being a cause to an effect—there are some presuppositions on the ontological level. There is a shift in perspective from an essentialist view, according to which information is understood as a property of a thing, to a procedural view, according to which information is considered a result of a process. Oyama, who is founder of evolutionary developmental systems theory (DST), wants to solve the gene–environment dualism. She therefore starts with the concept of interaction, which presupposes the ontological existence of two independent parts (interactants).

Oyama refers to the object of evolution. In the Darwinian tradition, we focus on the evolution of species. In it, the environment has an effect on genes (by natural selection), but from an ontological perspective both are treated as *in*dependent entities. In contrast, DST is based on reciprocal dependence, according to which environment and genes necessarily evolve together in developmental systems whose parts are *inter*dependent. They can be separated conceptually, but since concepts are always man-made one errs if genes and environment are seen as independent entities. From this point of view, the consequence of this gene–environment dualism is always one-sided determinism. This is because the question of first causality only arises on the basis of a dualistic presupposition and its answer necessarily requires the reduction of one side.

Yet what is the status of genes and environment for DST? They are considered a resource, from which semantic information is generated in the developmental process. It is important to realize that the phenomena that make GD plausible can consistently also be explained from a resource perspective. When certain gene products are not present, either because the encoding genes do not exist (pieces loss), are mutated (dysfunctional protein), or are epigenetically deactivated, they cannot serve the development system as resources, with the consequence that the development of the relevant character deviates from the norm. Such deviations can also be caused by supernumerary genes (e.g., trisomy 21) or by a similar imbalance on the environmental side (e.g., water or nutritional deficiency). It is crucial to see that the phenotypic differences are not determined by preexisting instructions, but are caused by a change that affects the whole developmental system. Therefore the concept of the *resource* could and should replace the overloaded concept of *information*.

Although this reason for GD has been known since decades, Oyama's solution has not prevailed in the public discourse on genetics, which seems to speak for Kitcher's thesis of oversimplification. But in the scientific discourse, Oyama's approach has had a formative impact. Because of this argumentation, the resulting viewpoint is also referred to as *molecular epigenesis* (see Burian 2004; Stotz 2006): Organisms emerge systemically in a reciprocal construction process, in which both genes and environment are of equal importance. It is important finally to remember Waddington's initial intention since the original synthesis concept of epi-genetics refers to the same constructional view.

References

Bartram, C. R., Beckmann, F. B., Breyer, F., Fey, G., Fonatsch, C., Irrgang, et al. (2000). Probleme genetischer Determiniertheit. In C. R. Bartram, F. B. Beckmann, F. Breyer, G. Fey, C. Fonatsch, B. Irrgang, et al. (Eds.), *Humangenetische Diagnostik. Wissenschaftliche Grundlagen und gesellschaftliche Konsequenzen* (pp. 5–50). Berlin: Springer.

Burian, R. M. (2004). Molecular epigenesis, molecular pleiotropy, and molecular gene definitions. *History and Philosophy of the Life Sciences, 26,* 59–80.

Cooney, C. A. (2007). Epigenetics—DNA-based mirror of our environment? *Disease Markers, 23,* 121–137.

Crick, F. (1958). On protein synthesis. *Symposia of the Society for Experimental Biology, 7,* 139–163.

Der Spiegel. (2010). Der Sieg über die Gene. Klüger, gesünder, glücklicher: Wie wir unser Erbgut überlisten können. *Der Spiegel, 32,* 1.

Fischer, A. (2013). Die Epigenetik neurodegenerativer Erkrankungen. *Spektrum Wiss, 7,* 30–38.

Fraga, M. F., Ballestar, E., Paz, M. F., Ropero, S., Setien, F., Ballestar, M. L., et al. (2005). Epigenetic differences arise during the lifetime of monozygotic twins. *Proceedings of the National Academy of Sciences of the United States of America, 102,* 10604–10609.

Galton, F. (1875). The history of twins as a criterion of the relative powers of nature and nurture. *Journal of the Anthropological Institute of Great Britain and Ireland, 5,* 391–406.

Godfrey, K. M., Lillycrop, K. A., Burdge, G. C., Gluckman, P. D., & Hanson, M. A. (2007). Epigenetic mechanisms and the mismatch concept of the developmental origins of health and disease. *Pediatric Research, 61,* 5R–10R.

Griffith, P. E. (2006). The fearless vampire conservator: Philip Kitcher, genetic determinism, and the informational gene. In E. Neumann-Held & C. Rehmann-Sutter (Eds.), *Genes in development: Re-reading the molecular paradigm* (pp. 175–198). Durham: Duke University Press.

Griffith, P. E., & Knight, R. D. (1998). What is the developmentalist challenge? *Philosophy of Science, 65,* 253–258.

Griffith, P. E., & Stotz, K. (2013). *Genetics and philosophy—An introduction.* Cambridge: Cambridge University Press.

Holliday, R. (1990). Mechanisms for the control of gene activity during development. *Biological Reviews of the Cambridge Philosophical Society, 4,* 431–471.

Jablonka, E., & Lamb, M. J. (2002). The changing concept of epigenetics. *Annals of the New York Academy of Sciences, 981,* 82–96.

Kitcher, P. (2001). Battling the undead: How (and how not) to resist genetic determinism. In R. Singh, K. Krimbas, D. Paul, & J. Beatty (Eds.), *Thinking about evolution: Historical, philosophical and political perspectives* (pp. 396–414). Cambridge: Cambridge University Press.

Lemke, T. (2002). Mutationen des Gendiskurses. Der genetische Determinismus nach dem Humangenomprojekt. *Leviathan: Berliner Zeitschrift für Sozialwissenschaft, 30,* 400–425.

Mahner, M., & Bunge, M. A. (2000). *Philosophische grundlagen der biologie*. Berlin: Springer.

Markert, D. (2008). *Das Jungbrunnenwunder. Der Markert-Plan für 120 Jahre Lebenskraft*. Hannover: Schlütersche.

Midgley, M. (1984). Reductivism, fatalism and sociobiology. *Journal of Applied Philosophy, 1*, 107–114.

Oyama, S. (1985). *The ontogeny of information: Developmental systems and evolution*. Cambridge: Cambridge University Press.

Riggs, A. D., Russo, V. E. A., & Martienssen, R. A. (1996). *Epigenetic mechanisms of gene regulation*. Plainview: Cold Spring Harbor.

Schuol, S. (2014). Kritik der Eigenverantwortung: Die Epigenetik im öffentlichen Präventionsdiskurs zum Metabolischen Syndrom. In V. Lux & T. Richter (Eds.), *Vererbt, codiert, übertragen: Kulturen der Epigenetik* (pp. 271–282). Berlin: De Gruyter.

Staege, B. (2014). Was ist Epigenetik? http://www.biomed-staege.de/html/epigenetik.html. Accessed May 8, 2016.

Sterelny, K., & Griffith, P. E. (1999). *Sex and death: An introduction to the philosophy of biology*. Chicago: University of Chicago Press.

Stotz, K. (2006). Molecular epigenesis: Distributed specificity as a break in the central dogma. *History and Philosophy of the Life Sciences, 28*, 533–548.

Tappeser, B., & Hoffmann, A.-K. (2006). Das überholte Paradigma der Gentechnik. Zum zentralen Dogma der Molekularbiologie fünfzig Jahre nach der Entdeckung der DNA-Struktur. *Umwelt, Medizin, Gesellschaft, 19*, 17–22.

Thomas, W. I., & Thomas, D. S. (1928). *The child in America: Behavior problems and programs*. New York: Knopf.

van Speybroeck, L. (2002). From epigenesis to epigenetics—The case of C. H, Waddington. *Annals of the New York Academy of Sciences, 981*, 61–81.

Vivamus. (2014). Anti-aging via epigenetik. Wissenschaftliche Hintergründe von age LOC und R2. http://www.vivamus-consulting.de/index.php?kat=34. Accessed May 8, 2016.

Waddington, C. H. (1952). The evolution of developmental systems. In D. A. Herbert (Ed.), *Proceedings of the Twenty-Eighth Meeting of the Australian and New Zealand Association for the Advancement of Science* (pp. 155–159). Brisbane: A.H Tucker Government Printer.

Waddington, C. H. (1957). *The Strategy of the genes: A discussion of some aspects of theoretical biology*. London: Allen and Unwin.

Waddington, C. H. (1968). The basic ideas of biology. In Ders. (Eds.), *Towards a Theoretical Biology. An IUBS symposium* (pp. 1–32). Edinburgh: Edinburgh University Press.

Youngson, N. A., & Whitelaw, E. (2008). Transgenerational epigenetic effects. *Annual Review of Genomics and Human Genetics, 9*, 233–257.

Author Biography

Sebastian Schuol Dr. is scientific coordinator of the EURAT-project (Ethical and Legal Aspects of Whole Genome Sequencing) which is located at the University of Heidelberg. After studying philosophy and molecular genetics he was a fellow at the DFG Research Training Group "Bioethik" at the International Center for Ethics in the Sciences and Humanities (IZEW) in Tübingen and completed his dissertation in philosophy of biology on the theoretical and practical impact of epigenetics on the concept of the gene. He has a special interest in philosophy of science, theory of biology and bioethics. So far he has published two articles on epigenetics (Kritik der Eigenverantwortung: Die Epigenetik im öffentlichen Präventionsdiskurs zum Metabolischen Syndrom, Berlin 2014; Der Lebensstil als Biotechnik? Zur Erweiterung des Genbegriffs durch die Epigenetik, Tübingen 2015).

Contact: National Center for Tumor Diseases (NCT) Heidelberg, Im Neuenheimer Feld 460, 69120 Heidelberg, Germany.

Identity and Non-identity. Intergenerational Justice as a Topic of an Ethics of Epigenetics

Philipp Bode

Abstract

Obviously some effects of epigenetics are organized in a paradox way. We are aware of this since the famous Överkalix Studies at the latest. It became obvious that nutrition during the childhood of the father's father significantly influenced the life expectancy of grandchildren: a rich diet between the 9th and 12th year of life, the slow growth period, significantly increased the grandchildren's risk of catching diabetes and die. A poor diet, on the other hand, seemed to reduce the risk of their sons suffering from a cardiovascular disease. May we expect our child to put up with a – somewhat – poor diet, as long as this way a certain (and possibly not even completely grasped) general mortality risk of this child's potential grandchildren might be statistically reduced? Do parents, when nourishing their child, even have a moral obligation of considering a risk of disease for a potential generation of great-grandchildren? Or should the parents' view be exclusively on raising their own children which, however, would be a serious blow for general intergenerational justice? It seems to be immediately wrong to prevent one's own children from a healthy diet only for the sake of possible great-grandchildren. However, ethics must be capable of explaining *why* this is wrong. Thus, by three steps I will attempt to demonstrate that this instinctive rejection is indeed justified.

P. Bode (✉)
Institut für Philosophie, Leibniz Universität Hannover, Im Moore 21 (Hinterhaus), 30167 Hannover, Germany
e-mail: philipp.bodephilos@uni-hannover.de

© Springer Fachmedien Wiesbaden GmbH 2017
R. Heil et al. (eds.), *Epigenetics*, Technikzukünfte, Wissenschaft und Gesellschaft / Futures of Technology, Science and Society,
DOI 10.1007/978-3-658-14460-9_5

55

1   Introduction

For the time being there is no more or less substantial ethics of epigenetics. It may be assumed that an ethics of epigenetics is not believed to be necessary because the issues to be tackled are similar to those of genetic research or are so close to the latter that both lines of ethical argumentation can be developed parallel. However, this is not always the case. Indeed, epigenetics creates only a few *specifically* moral problems—but still it does create problems which should be taken seriously and in some cases can be dealt with by help of practiced ethical-argumentative tools. In the following I would like to demonstrate this by the example of one concrete aspect. To do so, after having introduced the basic situation, I will make a longer detour which seems to leave the track of epigenetics but finally comes back. In the following it will not be about an attempt to outline an ethics of epigenetics as such but about showing which specific problems may be created by epigenetic insights.

2   A Moral Paradox?

Obviously some effects of epigenetics are organized in a paradox way. We are aware of this since the famous Överkalix Studies (Kaati et al. 2002) at the latest. In the course of several extensive studies based on an ever bigger genealogic and medical-historic set of data from the North-Swedish province of Överkalix, between 1999 and 2007 the researchers were able to demonstrate that a proneness to diabetes as well as cardiovascular diseases in connection to the nutrition of the generation of grandparents are gender-specifically inherited. Accordingly, it became obvious that nutrition during the childhood of the father's father significantly influenced the life expectancy of grandchildren: obviously a rich diet between the 9th and 12th year of life, the slow growth period, significantly increased the grandchildren's risk of catching diabetes and die. Obviously a poor diet, on the other hand, seemed to reduce the risk of their sons suffering from a cardiovascular disease. Over the time these results were extended insofar as not only the father's father could epigenetically affect the generation of the grandchildren but also his wife, the mother of the father generation. However, this effect seemed to be strictly gender-specific: obviously the nutrition of the grandfathers had exclusively epigenetic effects on the male grandchildren, that of the mothers' mothers exclusively affected granddaughters. The entire mother's side showed no significance for these studies.

All these studies were repeated in Bristol in England, and their results were verified by way of a much bigger random sample (n = 14,000) (Pembrey and

ALSPAC Study Team 2004; Pembrey 2002). In this context it must be taken into consideration that much of the responsibility is not with *one's own* conscious way of life but with the way in which the parents organized the way of life of their children. If accordingly—roughly speaking—the first 12 years (including the time of pregnancy) are crucial for epigenetic effects on the follow-up generation, it is the parents who must behave responsibly. If, however, the latter are confronted with the insight that a poor diet might affect their potential great-grandchildren, and that is in the *positive* sense (by minimising a certain risk of disease), this results in a paradox situation of decision-making. May we expect our child to put up with a—somewhat —poor diet, as long as this way a certain (and possibly not even completely grasped) general mortality risk of this child's potential grandchildren might be statistically reduced? Do parents, when nourishing their child, even have a moral obligation of considering a risk of disease for a potential generation of great-grandchildren? Or should the parents' view be exclusively on raising their own children which, however, would be a serious blow for general intergenerational justice?

It seems to be immediately wrong to prevent one's own children from a healthy diet only for the sake of possible great-grandchildren. However, ethics must be capable of explaining *why* this is wrong. Thus, by three steps I will attempt to demonstrate that this instinctive rejection is indeed justified. By a first step, the question of where this conflict comes from must be clarified and if it may claim validity at all. This means: are their individuals who, in retrospect, *may* morally accuse me (may it be my children or my great-grandchildren)? I would like to demonstrate that—contrary to some arguments of bio-ethics—this is indeed the case. By a second step it must be clarified *who* is in principle capable of such an accusation. A third step is supposed to show why in this particular case I will *not* necessarily be morally accused although there might be potential accusers.

Of course the Överkalix Study is based on genealogic data and not on molecular-biological ones. If, however, the suspicion of intergenerational epigenetic inheritance can be substantiated, ethical questions of intergenerational justice will be raised.

3　Is a Retrospective Moral Accusation Possible?

From embryonic research we know the argument that currently existing beings must always get preference over potentially existing beings. Just recently, Anja Karnein in her study *A Theory of Unborn Life* (Karnein 2012) has supported the thesis that in certain cases currently existing individuals have the right to interfere with the genomes of future persons. The crucial question in this context is how to deal with embryos which will develop into individuals to make sure that the individuals into whom they develop will not suffer *injustice* from our behaviour. In my opinion it is possible by way of this argument to develop a preliminary approach at a solution also for the question if we are allowed, as far as health is concerned, to give currently existing individuals preference over future individuals (thus consciously interfering with their epigenomes) even if this might result in disadvantages for these future individuals.

4   Karnein's Theses

Individuals whose dignity is protected, says Karnein, are all those who are currently or have been capable of acting morally. Among them there count newborn babies, that is human beings who have never been capable of acting morally, are currently not capable and might never be capable. The concept of the person is based on the social world of those being capable of acting morally, and newborn babies as well as disabled or coma patients are part of this world. However, this extended concept of the person does not include embryos. For Karnein, what is crucial for this is the fact that unborn human beings are connected to another, irreplaceable organism, that is that of the pregnant woman. Karnein is of the opinion that every pregnant woman has the right to withdraw her support of the foetus growing in her body and thus also accept its death. This is the case because the concept of the person is not only meant for the group of human beings who are not dependent on life-sustaining measures, furthermore they are also supposed to be *independent of another existence*, such as the womb. For Karnein, a foetus becomes a person only by its birth.

From this thesis there concludes that *born* human beings have a claim to moral protection, whereas this does not hold for *unborn* human beings.

Although against this background unborn life has no claim to moral protection, nevertheless—as is Karnein's second thesis—embryos must be treated *as if* developing into persons, and that is as long as it is not clear that they will *not* become persons. This means that (almost[1]) all kinds of artificial interventions and inflictions of damage are unacceptable as long as they still *may* become persons (that is as long as they have not yet been killed). Thus, if an embryo's future is not clear—and this concerns the majority of cases—its epigenetic features must not be

---

[1]Exceptions are all those interventions as are legal with adults without their permission.

manipulated. This obligation holds also in retrospect: individuals may claim that the embryos out of which they have developed have *not* been manipulated.

This means that unborn human beings, although not morally protected, must be protected from damaging manipulation as long as it is unclear that they will *not* develop into persons. This holds already because we cannot always know if these embryos will actually develop into persons, and if, we do not know which one. Furthermore, it concludes from this argument that embryos may be killed because then they will not develop into persons, so that there will be no harm for future individuals. According to this concept, there is no obligation to allow all embryos to develop into persons.

However Karnein—and we—are not primarily interested in the question if we are allowed to kill embryos but if we may manipulatingly interfere with their epigenomes, that is if we may change the embryo's 'natural equipment'. To answer this question, it was indispensable to develop a concept of the person which includes both current and future persons in the same way.

5   Identity, Non-identity and Intergenerational Obligations

The question of the legality of embryonic manipulation—just like some essential aspects of an ethics of epigenetics—touches a problem of *intergenerational obligation*. Intergenerational obligations raise two fundamental theoretical issues. The first one is the so called *non-identity problem*. As far as by *damaging* we mean causing a condition under which persons are worse off than before, we are not capable of doing harm to future persons, as for them there has not been any time *before*. Thus, before their existence (that is before conception) we are not capable of doing harm to future persons. The second problem—the *identity problem*—is that obviously even *after* conception we are incapable of doing so, as obviously—at least from a first person point of view—it would be unreasonable to claim that somebody's situation has been made worse than a theoretical alternative situation in which he/she would be, had his/her embryonic development been different. The reason for this is simple enough: had the embryonic development been different, somebody else would have been born. Postponing pregnancy does not mean having the same child, only with different features, at a later time, but having a *different* child (Parfit 2006). And it seems that also this argument is retrospectively valid: obviously individuals may hardly claim that they have been harmed, after all, the alternative would be their non-existence or the existence of somebody else.

Thus, the problem of non-identity only weakly concerns the ethics of epigenetics, and only if the question concerns an alternative version of one's own self. Karnein goes on discussing this point, now as the *problem of identity*, which definitely makes sense conceptually. By reference to David Velleman, it is not only unreasonable to include those into the argument who might have been born instead of ourselves, in Velleman's eyes it is as unreasonable to, from the first person point of view, take possible alternative versions of ourselves into consideration. Had things gone different in our past, today we would be different persons, and all measures applied *to* us *before* our birth would be part of *our* biographies, and without these measures *we* would not be *ourselves.*[2]

Now it might be that also the knowledge of epigenetic effects falls under this category. Should a certain way of life affect a child's quality of life, still it would be this particular child which would be born and, due to its birth, would be given preference, which is why any objection towards the parents would be unreasonable in view of its moral content.

6   Objections Against Karnein

Karnein's argument shows much plausibility, however it is based on at least one problematic assumption. This is that *for the child*, from a first person point of view, being born is *intrinsically good* and must be welcomed, that thus having preferred it from any other child may relieve the parents from certain kinds of guilt. Probably this assumption has not been verified because it is convincing due to its urgent plausibility, thus looking axiomatic. Putting this axiom into question seems to imply some dark denial of life, thus diminishing any justification to comment at all on a question of life. But is the moral value of being born really unquestionable? Of course it is impossible to judge from a perspective of non-existence, thus one cannot judge positively on one's own non-existence. The idea of one's own non-existence (as far as it is possible at all) is necessarily connected to one's own existence. Nevertheless, nothing speaks against considering one's own existence worthless and not worthy of being experienced. It is logically consistent that we can imagine a world without us only from a world with us; it is a fact, however, that still we leave this world according to our own wish and by our own hands. Thus, at first nothing speaks against the assumption that one's own non-existence is desirable.

---

[2]In the US TV series *Fringe* this thought experiment has been acted out, and an in a surprisingly compelling way.

7 One's Own Existence as a Harm

In the English-speaking countries there has been a lively debate on this since the 1980s, taking the assumption that having been born can indeed be considered harm very seriously. The objection that according to this theory there is no possibility to assume that there is no harm, as there is no *before* for the harmed individuals, is correct, but obviously only from a third person point of view. Consequently, such theories are limited to the question if there is a moral obligation of *conception*, thus they judge from a hypothetical parents point of view (Govier 1983; Vetter 1969, pp. 445–447; Vetter 1971, pp. 301–302). The first person point of view, which indeed seems to be problematic, was assessed e.g. by the US American moral philosopher Jefferson McMahan who, surprisingly, came to the conclusion that from the point of view of existing individuals it is precisely the fact of possible *non*-existence which justifies an objection (McMahan 1986)—thus threatening to clash with Karnein's thesis, as she had argued that an objection by an existing person is implausible because the alternative would be one's own non-existence, and nobody is likely to want this.

According to McMahan's thesis, it is morally required to safeguard the existence of future persons. This requirement results not only from keeping the rights and interests of future persons, but the reason is impersonal. It would have to be argued that there is a principle of avoiding harm, according to which we should help humans—for their own sake—to existence only if there is the probability that they will live a life worth 'living'. Behind this thesis is the assumption that there is indeed a life not worth living and that there is *no* requirement to under *all* circumstances prefer living over non-existing. Helping a human to existence whose life is obviously not worth living means deliberately *harming* this person, and there is the plausible moral requirement of not harming other people.

If, however, bringing a human to life while knowing that he/she faces a life *not worth living* means *harming* this person, to a certain degree somebody would be *treated preferentially* if he/she was born while we know that he/she will face a life *worth living*. Then, being born would always be a kind of preferential treatment. Consequently, when comparing these two principles, there seems to be the requirement to help humans to existence only as far as their life seems to be worth living. Although it seems to be correct that humans, once they have been born, are treated somewhat preferentially, it is wrong to derive *from this* a moral *obligation* to actually help them to existence. The reason is already known: preventing future generations or individuals from existing violates neither their interests nor their rights because the rights or interests of non-existing persons *cannot* be violated.

Consequently, there is no moral obligation to help future humans to existence. Only if these future persons *will actually* exist, we have obligations towards them, and that is the obligation to make their lives worth living. Bringing a human to existence under the premise that his/her life will be worth living (whatever that means exactly) is, as already mentioned, not an obligation per se, nor is it required to take care that the evil in the world will not spread further. For McMahan, there is an obligation only towards the individual we have already helped to existence.

McMahan calls this argument the 'objection condition'. It means that an objection is only valid if there is somebody to make it. Under these conditions, a person who has been brought to a life not worth living may blame his/her parents for having made this existence possible. This is—notwithstanding the problematic nature of McMahan's text—the crucial result for our context: not any life is in principle better than no life, under certain conditions non-existence is an option.

Together with McMahan I would like to object against Karnein that for a child (or the later person) being born is *not* always intrinsically good and must be preferred to non-existence. McMahan supports the thesis that there are at least some *special* cases in which an objection against the parents, about one's own existence, is justified, that is those cases in which one's own life is considered not worth living.

Here at the latest the problem comes to the fore of what, after all, we mean by 'not worth living' and who is to decide about this status. According to McMahan's argument, it is the parents who decide about the status for their children, by not being morally obliged to help a child to existence. However, what if the born person him/herself considers his/her life not worth living, thus contradicting the parents' decision? Both ways of arguing, McMahan's and Karnein's, are forced to introduce a limiting value which defines from when on there is a moral obligation. In view of the question of the permissibility of embryonic manipulation, Karnein introduces a limiting value below which, in view of the conception of a child, we may not fall and which consists of the fact that the future person's *independence* must be guaranteed. For McMahan, a *life worth living* is such a limiting value. As it is certainly a necessary fact that such limits cannot be defined by persons not yet existing, these limits may only claim a limited stability as long as they are defined by existing persons alone.

This is meant to say: individuals, once they have been born, must be capable of co-deciding about these limits—in my opinion this is exactly what is implied by Karnein's demand for independence. Even if we assume that parents have decided, to the best of their knowledge and belief, that their offspring does *not* fall below an appropriate limiting value of independence or a life worth living, the case must be taken into consideration that the individuals themselves, once they have been born, get the impression to have fallen below this limiting value. Thus, what would speak against an objection towards the parents solely referring to the fact of having been brought to existence by them?

8   The Subject of Retrospective Moral Accusation

Once again: For Karnein, objections against personal features (which may be the result of genetic manipulation) are not possible, as an objection against the actual ontogenetic process would be an objection against one's own existence. For McMahan, on the other hand, such objections are possible, at least if an individual is living a life not worth living, that is if the parents have not met their obligation to help their child to a life which is worth living (as far as they were in a position to influence this at all), thus if at least in this case there are reasons to object to one's own existence. Both authors need a limiting value for these restrictions of the possibility of retrospective objection. Both authors leave the decision about this limiting value to the parents, either exclusively (Karnein) or mostly (McMahan).

In my opinion, the thus resulting neglecting of the concerned individuals' right to have a say, i.e. of those persons about whose independence and living a life worth living we are speculating, is a mistake. *Any* individual has the right to retrospectively object against his/her parents because the decision about one's own life being worth living is only partly made by the parents and also partly by the *experiencing individual him/herself.* The actual preferential treatment of one's own existence is no argument against such a possibility to object because it assumes that preferential treatment has a *value.* However, the fact that I am here and not somebody else is not a value but indeed a fact. Neither did my parents intend to conceive precisely *me* (even if throughout my life they have told me different), nor is it that I am more *valuable* than any alternative version of my parents' child, as this version is non-existent. Consequently it is no value judgement that *I, of all,* exist, and thus I do not have to accept possible harm due to my parents' behaviour, only because the alternative would be my non-existence. I *may* be thankful for my existence, but I do *not* have to, as there has been preferential treatment of any alternative version of myself which may be *valuable as such.*

It is true: if I complain about certain features, and the alternative would be my non-existence, I would have to complain about my own existence. If it concerns marginalia, this may look unreasonable, however nothing speaks against the indeed even plausible case that, due to a grave illness or to certain features *which in my opinion are unacceptable* I accept this alternative. Nothing speaks against putting one's own existence into question or against the wish to make an end to one's own existence.

Individuals are in principle entitled to morally accuse people from the preceding generations for their behaviour as far as this behaviour has negatively affected the individual concerned, including also *epigenetic* effects. The argument that any accusation means putting one's own existence into question is not convincing.

However, this argument needs further substantiation. At first it is a fact that the arguments I have stated here understand our existence, and thus also my own existence, as being *ethically neutral* compared to any alternative version of ourselves. Of course this is not a matter of course position, possibly it is not only deeply untheological but also Kierkegaard and in particular Hans Jonas would vehemently disagree (Jonas 1984, Chap. 2, IV–V).[3] And of course the probably inevitable fact cannot be denied that the overwhelming majority of all existing humans have an interest in their existence, and it is also perfectly all right that usually we are more interested in our *own existence* than in any other real existing existences. However, this has got nothing to do with *me* having moral advantages to any *alternative* child of my parents. This is meant to say: there is no moral advantage of any existing to any non-existing person. From this there concludes: there is *no value* in *my* existence instead of any alternative version of my parents' child, not even if, from a first person point of view, I consider my own existence reasonable—perhaps even logical. Consequently, my own existence does not prevent me from a moral accusation (for it is insignificant if I do not exist at all or if somebody else exists instead of me).

Thus it becomes obvious that actually two questions are hiding behind one. The first question is: Am I responsible towards my great-grandchildren (even if it is only statistically suggested)? The second question is: Would my great-grandchildren be *entitled* at all to morally accuse me in retrospect?

---

[3]For Jonas it is out of the question that the continuous existence of both the individual and mankind as a whole is an ethical obligation and that, from a first person point of view, the life of any individual is reasonable, as any existence is supposed to exist, as a value-immanent existence—the alternative being complete non-existence. As it is well known, Jonas's argument is based on ontological premises which are indeed doubtful (such as the necessity that life agrees with itself or that there is no alternative to existence compared to non-existence, making the former perfect).

9   Solution and Prospect

As we have seen, I am much concerned with the in principle possibility of ret-rospective objections. However, does this hold also for our particular case? Let us try to prospectively imagine a moral accusation by our grandchildren, to decide about our current behaviour according to this accusation. Indeed, the existence of our great-grandchildren is not even given. They are not even future persons in the sense of, for Karnein, embryos being future persons, as not only there has not been any conception yet but possibly both parents are not yet alive to conceive these great-grandchildren. What sense does a scenario make in which one's own child has only limited access to a healthy diet if this child will never have offspring of its own, it possibly *my behaviour* prevents it from conception? Vice versa, I might witness my own child leading only a somewhat healthy life but not live to see my then immunostimulated great-grandchildren.

Whatever a responsibility towards my grandchildren may look like, it does *not* consist of malnutrition, mistreatment or bad education of my own children. I may be concerned by my grandchildren's accusation as far as I do something wrong immediately concerning them. For this purpose, they may themselves co-decide about their limiting values and may morally accuse me to a much bigger extent than allowed by Karnein and McMahan. However, I cannot be concerned if I have done something good, and providing a healthy diet for one's own children may not be supposed to be wrong in any culture. Thus, the answer to the first question is: I do *not* have a responsibility towards my grandchildren if my behaviour as such is morally correct. And providing a healthy diet, education etc. for my own children *is* good behaviour.[4]

The answer to the second question, if my grandchildren are entitled at all to hold me morally accountable as far as my child's diet has actually affected their immunostability, provides that this is a reasonable question at all. They would have to condemn an actually good action—providing a healthy diet for my children, that is their parents—to advance their claim. And even if we assume that they are entitled to such a verdict, their accusation would be part of an imaginary scenario whose existence, that is the probability of their own existence, is more improbable than that the embryo will develop into a future person, and we note that at the moment of our argument this has *not* yet happened. Judging in the sense of such an

---

[4]This argument may also be supported by the principle of ethical ambivalence. According to this, a healthy diet for my children is the (morally permissible) man action, my grandchildren's statistically weakened immunostatus is the (unintended and thus accepted) side effect.

imagination, while not doing a real (and morally permissible) deed (providing a healthy diet for children), is itself morally unacceptable. Thus, the answer to the second question is: my grandchildren are *not* entitled to morally accuse me in retrospect, at least as long as to do so they would have to condemn a behaviour which as such is morally permissible.

Thus ends the argument on intergenerational justice. At least I have proven that there is indeed the possibility to be morally accused in retrospect and that the number of individuals is basically not limited. As soon as this had been clarified, reasons could be developed why in our particular case still *no* moral accusation would be justified. Recently there has been much talk of transgenerational epigenetic inheritance (Grossniklaus et al. 2013; Daxinger and Whitelaw 2010; Jablonka and Raz 2009); it seems as if the concept of inheritance as a whole is being reformatted, at least partly. Even if the Överkalix Studies do not allow for *causally* concluding from the grandfathers' diet on their grandchildren, it will not do any harm if we take possible ethical tensions out of obviously paradoxically organised epigenetic effects without being in the need of finally judging on the stability of the assumption of epigenetic inheritance.

References

Daxinger, L., & Whitelaw, E. (2010). Transgenerational epigenetic inheritance: More questions than answers. *Genome Research, 20*, 1623–1628.

Govier, T. (1983). What should we do about future people? In J. Narveson (Ed.), *Moral issues* (pp. 55–74). Oxford: Oxford University Press.

Grossniklaus, U., Kelly, B., Ferguson-Smith, A. C., Pembrey, M., & Lindquist, S. (2013). Transgenerational epigenetic inheritance: How important is it? *Nature Review Genetics, 14*(3), 228–235.

Jablonka, E., & Raz, G. (2009). Transgenerational epigenetic inheritance: Prevalence, mechanisms, and implications for the study of heredity and evolution. *The Quarterly Review of Biology, 84*(2), 131–176.

Jonas, H. (1984). *Das Prinzip Verantwortung. Versuch einer Ethik für die technologische Zivilisation*. Frankfurt am Main: Suhrkamp.

Kaati, G., Bygren, L. O., & Edvinsson, S. (2002). Cardiovascular and diabetes mortality determined by nutrition during parents' and grand parents' slow groth period.*European Journal of Human Genetics, 10*, 682–688.

Karnein, A. (2012). *A theory of unborn life. From abortion to genetic manipulation*. Oxford: Oxford University Press.

McMahan, J. (1986). Nuclear deterrence and future generations. In Avner Cohen & Steven Lee (Eds.), *Nuclear weapons and the future of humanity* (pp. 319–339). Totowa, NJ: Rowman and Allanheld.

Parfit, D. (2006). Rights, interests and possible people. In H. Kuhse, & P. Singer (Eds.), *Bioethics. An anthology.* (2nd Ed., pp. 108–112) Oxford: Blackwell Publishing.

Pembrey, M. (2002). Time to take epigenetics inheritance seriously. *European Journal of Human Genetics, 10,* 669–671.

Pembrey, M., & ALSPAC Study Team. (2004). The Avon Longitudinal Study of Parents and Children (ALSPAC): A resource for genetic epidemiology. *European Journal of Endocrinology, 151,* 125–129.

Vetter, H. (1969). The production of children as a problem of utilitarian ethics. *Inquiry, 12,* 445–447.

Vetter, H. (1971). Utilitarianism and new generations. *Mind, 80,* 301–302.

Author Biography

Philipp Bode is research fellow at the Centre for Ethics and Law in the Life Sciences (CELLS) and assistant lecturer at the Institute of Philosophy, Leibniz University Hannover. He is the initiator of the working group "Ethics in Epigenetics" at the Akademie für Ethik in der Medizin. His research interests include ethics, applied ethics (especially bioethics, animal ethics, media ethics), philosophy of mind, philosophical anthropology and philosophy and history of sciences. Some of his publications include Bode, S.; Murawski, C.; Soon, C.S.; Bode, P.; Stahl, J.; Smith, P.L. (2014). Demystifying "free will": The role of contextual information and evidence accumulation for predictive brain activity. Neuroscience & Biobehavioral Reviews 47, 636–645. Bode, P. (2015). An den Grenzen der Vernunft. Eine kurze Geschichte der Gottesbeweise. In Nickl, P. & Verrone, A. (Hrsg.), Wie viel Vernunft braucht der Mensch? Texte zum 3. Festival der Philosophie (S. 139–165). Münster: Lit Verlag.

Contact: Institut für Philosophie, Leibniz Universität Hannover, Im Moore 21 (Hinterhaus), 30167 Hannover.

Genetics, Epigenetics and Forms of Action. About the Ethical Ambivalence of Epigenetic Knowledge

Joachim Boldt

Abstract

Epigenetics can be considered as a continuation of genetics. Even if it includes environmental factors in its analyses, the focus remains on the cellular-molecular perspective on the organism. However, epigenetics expands the spectrum of what is known from the discussion about ethical aspects of genetics in two regards. First of all, the inclusion of environmental factors causes the relationship of the individual that is provided with predictive diagnoses to change with this knowledge. Secondly, this correlation leads to the question as to how the instrumentally useful epigenetic findings for the patient can be embedded in the applicable context of social-communicative action that is relevant for the patient's everyday life.

1   Introduction

"Epigenetics demonstrates that humans are not a constant" (Mansuy 2014): Those who work with epigenetics frequently come across such and similar statements. Epigenetic research is surrounded by an aura of liberation of humans from a genetic-deterministic stranglehold (cf. Schuol in this publication). How can this aura be explained, can it be justified, and which ethical questions does epigenetics ultimately raise?

J. Boldt (✉)
Institut für Ethik und Geschichte der Medizin, Albert-Ludwigs-Universität Freiburg,
Stefan-Meier-Str. 26, 79104 Freiburg, Germany
e-mail: boldt@egm.uni-freiburg.de

© Springer Fachmedien Wiesbaden GmbH 2017   69
R. Heil et al. (eds.), *Epigenetics*, Technikzukünfte, Wissenschaft und
Gesellschaft / Futures of Technology, Science and Society,
DOI 10.1007/978-3-658-14460-9_6

By its nature, epigenetics belongs to the fields of molecular genetic research and technology. It does not deal with the function of the genes itself, but rather with the structures of the genome that can be acquired or changed during the course of an organism's life. However, these structures are also molecular-genetic structures that play a regulating role in determining genetic activity. Accordingly, like genetic characteristics, they can contribute to the molecular-genetic explanation of the origins of diseases, and epigenetic findings can be used as an approach for molecular-genetic forms of therapy.

Still, epigenetics is known as a field of research that represents an alternative to genetic research programmes. In the old discussion as to whether nature ("nature" in the form of deoxyribonucleic acid, DNA) defines the characteristics and abilities of organisms, or whether behaviour and environmental influences ("nurture") play the decisive role in the development of characteristics and abilities, epigenetic research findings are often interpreted as proof for the latter hypothesis. It then seems like epigenetics is caught between two stools. On the one hand, it is molecular genetics, and on the other, it seems to open the door for explanatory approaches to the behaviour and characteristics of organisms, which go beyond the molecular-genetic forms of explanation and point towards behaviour and environment. Now how can this position of epigenetic research between molecular genetics and organism-environment interaction be described and understood more accurately and what is the possible ethical relevance of this intermediate position?

2 Epigenetics and Genetics

According to its object of research, epigenetics is a molecular-biological science that deals with the mechanisms of gene regulation and gene expression. In doing so, it does not deal with factors of gene regulation and expression that are anchored at the DNA level, and like DNA altogether, are only changed through mutation. In contrast, epigenetics analyses regulation mechanisms that act on a "superordinate" level, e.g. the chromatin packaging of DNA, and which can be added or triggered by environmental influences (Youngson and Whitelaw 2008; Cremer 2010, pp. 51ff.). An important form of this chromatin modification is DNA methylation, where a methyl group is bound to the base cytosine. When looking at the entire genome, it results in a characteristic methylation pattern in the DNA (cf. Walter in this publication).

Genes that are silenced by methylation can no longer be read and are therefore inactive. Methylations can be modified by environmental influences, which include malnutrition and methyl group-rich nutrition as well as the brood care intensity.

Although methylations are always reversible, they can remain stable for many years if there are no corresponding reversing environmental factors. Furthermore, the methylation patterns are passed on when somatic cells divide, and there are indications that they are heritable to subsequent generations through the germ cells. However, uncertainty remains as to how this transgenerational heredity takes place on a molecular-genetic level (Cremer 2010, p. 63).

The methylation patterns play a role in the development of disease. For example, they are considered to be involved in the occurrence of diabetes mellitus type 2 and the susceptibility to stress. Therefore, epigenetics provides approaches for new therapies and new prevention measures. On the one hand, these therapies and measures can act at a molecular level and directly modify the methylation patterns. On the other, they could also consist of exposing the patients to certain environmental factors or to prevent their contact with such factors, in order to influence the patient's epigenome itself and possibly also that of his progeny. With regards to this latter possibility, a document from the German Federal Environment Agency states that:

> If mechanisms leading to health damage are known, disease prevention can be exercised on the one hand by avoiding harmful influences, and on the other, epigenetic regulation could be affected by dietary measures in the future [...] and the damaged caused by environmental influences could be repaired. (Süring 2010, p. 11)

For both cases, i.e. direct molecular-genetic intervention and the influence of environmental exposures, the purpose is to change specific, previously identified epigenetic structures that are involved in the development of disease. Although the field of epigenetics is interested in research on environmental factors and behaviour in the environment as causes for pathological epigenetic changes, possible therapeutic interventions always first aim to specifically modify these molecular structures.

With this orientation of epigenetics, the primacy of the field seems to remain clearly on the side of molecular biology and therefore the "nature" aspect in the nature-nurture debate. How is it then that some consider epigenetics as a pioneer for the revitalisation of the nurture perspective?

3   Epigenetics and Environment

The special role of the environment in the scope of epigenetic mechanisms is not already expressed in that individual environmental factors cause specific molecular changes in the cell nucleus, which then in turn affect ontogenesis and possibly phylogenesis. First of all, the existence of such mechanisms is not really biologically surprising, since, for example, DNA can also mutate because of environmental factors like radioactivity, which then in turn has an ontogenetic or also phylogenetic effect on the development of disease. Secondly, it is also not philosophically surprising, because it is commonly known that where and how we live has an effect on who we are and how we develop. It is not logically necessary that such effects can be transmitted through molecular biological mechanisms; however, if this is a fact, it seems at first glance that it does not contribute anything important to the significance of the link between environmental factors and individual development.

The tendency to claim epigenetics for the nurture side of the nature-nurture debate becomes understandable, if one considers the epigenetic proof of the effect of environmental factors on gene expression to be a principal change in the manner in which ontogenesis and phylogenesis are to be described. Indeed, epigenetics is often considered as a fundamental extension of classic evolution theory assumptions. The classic evolution theory assumes first of all that only the DNA itself is passed on from the parents to the progeny, and secondly, that changes in the DNA are a product of random mutations, which, on the long term, are passed on if they entail adaptive advantages. Now if epigenetic modifications of the genome can also be passed on transgenerationally, and if these modifications can be the result of the behaviour of an organism in its environment, then the heredity does not seem to be only a process of mutation and adaptability, but also of individual behaviour and the resulting epigenetic modification.

It seems that the direction of the cause-effect chain is fundamentally reversed here. While the classic evolution theory assumes that heredity can be explained by starting with the analysis of random mutations, and then looking at the resulting behaviours and their respective success, analysis in an epigenetic expanded theory begins with the individual behaviour, and then building onto this, investigates the epigenetic changes caused by this behaviour. Nicolosi and Ruivenkamp state accordingly that: "environmentally initiated novelties may have greater evolutionary potential than mutationally induced ones" (Nicolosi and Ruivenkamp 2012, p. 314).

Nicolosi and Ruivenkamp continue to argue that explanations starting at the level of the interaction between organisms and the environment effectively rehabilitate a series of terms and concepts that had been lost in post-Darwinism genetics and the evolution theory (cf. Seitz in this publication). They speak of an epistemological change in perspective in biology that is initiated by epigenetics and compare the older, "gene-centric" paradigm to the new, epigenetic paradigm. The gene-centric perspective describes ontogenesis and phylogenesis as a genetically determined process, where innovations only occur through random mutations (ibid., p. 313). In contrast to this, epigenetics would have to assume plasticity and flexibility of the genome and of the phenotype as well as creativity and contingency of the behaviour of organisms in their interaction with an environment (ibid., p. 317). Also for Canning, epigenetics fundamentally questions the genetic-deterministic paradigm of molecular biology. Contingency and dynamic interactions under the influence of many different factors should be considered instead (Canning 2008, p. 17).

From the classical evolution theory point of view, however, it can always be argued that the fact that environmental factors have heritable effects on gene expression does not mean that the basis of the explanation of genetic and evolutionary change can now be found in the interaction of organisms and their environment. Ultimately, behaviour in the environment can always be interpreted as a genetically controlled behaviour. Epigenetic influences of the environment may make genetic behaviour and development explanations more complex, but they do not make them impossible. In other words, taking account of epigenetic mechanisms does not necessarily mean that the individual, purposeful, learning, contingent behaviour of the organism in an environment now becomes the uncircumventable basis for the explanation of biological development processes. Thus, for example, one could postulate, like Cremer, that it is an adaptive advantage when the DNA of a genome is structured such that an epigenetically induced reaction to different nutritional situations is possible (Cremer 2010, pp. 64f.).

This argument shows that the answer to the question, whether behaviour explanations must be finally searched for at the genetic-molecular (or even atomic) structure level or whether they have to start with the individual and his interaction with other individuals and the environment, cannot be answered using empirical individual results. Such individual results can always be interpreted within the scope of these two explanation paradigms. The criteria of inner consistence and scope of explanation that can be formulated philosophically and epistemologically are more decisive for an answer to this question. Good arguments can surely be found there to grant the organism-environment interaction an explanation primacy (Boldt 2013).

However, it is important to remember that knowledge of the epigenetic mechanisms itself does not make this change inevitable in the scope of the concept of biological explanations. Although it might be assumed that a scientific cultural atmosphere is created with epigenetics, in which it is more widely acceptable to speculate about the fundamental significance of the influence of environmental factors on ontogenetic and phylogenetic development processes. Ultimately, the particular focus of epigenetics on the molecular-genetic processes demonstrates that it is only then really given weight in the scope of their organism-environment interactions when the corresponding intermediary epigenetic-molecular mechanism is identified.

4 Ethics of Epigenetics

In view of this philosophical-epistemological background, the ethical questions that are posed by epigenetics can now be analysed. If one first looks at the molecular biology side of epigenetics, it is then a science that deals with heritable molecular structures in the cell nuclei that affect the transcription and translation of DNA without modifying the DNA sequence itself. In this respect, it can be expected that the ethical questions of epigenetics would be similar to those of genetics.

If one limits oneself here to human aspects, a distinction can first be made between questions that involve genetic diagnostics and those involving interventions in the human genome. These interventions include in particular genetic therapy and research on therapeutic products that modify the DNA, but also changes to the DNA that aim to enhance human abilities beyond a healthy level. Like for genetics, however, it is also true for epigenetics that the therapeutic interventions in the epigenome, with few exceptions, are future visions that have not yet been redeemed.

For this reason, questions that involve genetic diagnostics, and also in the foreseeable future epigenetic diagnostics, are of greater current relevance. These questions are, listed without claim to be exhaustive: Is it helpful to know if one carries a genetic mutation that will surely lead to the occurrence of a serious disease later? Is it helpful to know with a certain degree of probability that a disease will develop later in life? Should relatives who could also be carriers of the mutation be informed of the results and be subjected to tests? Should the patient be informed of so-called incidental findings resulting from genetic investigations, without being the actual purpose of the investigation? Should genetic diagnostics be performed on germ cells?

All of these ethical questions regarding genetic diagnostics refer to the significance of genetic findings for individual persons, and therefore to aspects in which it is good or not for the individual to have this knowledge. The fact that these questions are posed relating to genetics is explained by the assumption that genetic findings enable particularly accurate predictions on the later occurrence of disease, and in more general terms, on the development of abilities and characteristics in a person. However, this assumption is not true or only partly true in many cases where the disease has multi-factorial origins. With monocausal genetic disorders such as Chorea Huntington, which usually only exhibits symptoms above an age of around 40, however, it becomes immediately obvious where the special challenge can be found in the handling of genetic knowledge.

Epigenetics increases the amount of data that needs to be recorded and considered when looking for explanations for the occurrence of disease on a molecular-genetic level. It is no longer sufficient to fully sequence the DNA; rather the epigenetically active parts of the genome packaging must also be recorded and set in relation to diseases. An obvious research approach along this line is to also strive for a sequencing of the epigenome after sequencing the human genome, and based on this data, to enable the most accurate predictions as possible on phenotypic developments, including the occurrence of diseases (Rivera and Ren 2013).

In this regard, epigenetics is an extension of genetics within the framework of molecular stratified medicine, which aims to correlate the development of disease in the most detailed manner possible with types of molecular, genetic and epigenetic structures (Rakyan et al. 2011). It can be assumed that the knowledge generated in this way, in turn, will mostly consist of probability statements on the later occurrence of disease, possibly with isolated exceptions, for which epigenetic effects are causally more clearly linked to certain diseases.

5 Handling of Epigenetic Findings

Essentially, epigenetics now adds an ethically relevant aspect to genetics. It concerns the relationship of individual patients to their respective epigenetic findings (cf. Fündling in this publication). Predictive genetic findings, in terms of a monocausal disease, are similar to the revelation of an irrevocable fate. As long as there is no genetic-therapeutic intervention that can prevent the development of the disease, the only option for the patient is to prepare for the inevitable. This aspect of genetic findings basically also remains if they indicate a probability of occurrence for a disease. Although the inevitable is then not the disease itself, but rather the stated probability, this probability is also a fact that needs to be dealt with.

In contrast, epigenetic findings potentially open the door to other behaviour options. Because also in extreme cases of predictive diagnosis of a monocausal epigenetic disease, it is first of all conceivable that these factors were caused by one's own behaviour or the behaviour of one's own ancestors, and second, because they could be reversed by one's own behaviour or the behaviour of one's own progeny, epigenetic findings can force one to assume responsibility. Epigenetic findings are knowledge that can potentially be reacted to with one's own actions, just as it is also possible to blame oneself or one's own ancestors (cf. Bode in this publication) that one's own epigenetic patterns are de facto what they are. In turn, this is also true for all those numerous cases where epigenetic findings indicate probabilities for the occurrence of a specific disease. These probabilities can also be changed by one's own behaviour, and therefore belong to the scope of individual responsibility.

This change from the level of epigenetic-molecular processes and laws to the level of a person's behaviour in an environment is special. As was already demonstrated, the fact epigenetics considers environmental factors to have a causal effect on molecular structures, which can then influence gene expression and are possibly transgenerationally heritable, does not yet mean that now the behaviour of an organism in its environment is to be considered as creative and contingent trial and error, which obliterates the boundaries of genetic determinism. Similarly, the fact that one's own epigenome can be changed by behaviour does not yet mean that this behaviour as such should be considered as valuable and important.

Those who interact with their environment and other people do so for the most varied of motives and with many different intentions. These include instrumental actions, which are characterised by the fact that they are performed when it can be expected that this action can help to achieve a very specific goal. However, interactions also take place when, for example, one interacts with others to attempt to solve a problematic situation, or when daily challenges are faced together. These actions do not have a previously defined and pursued goal, because it is unclear whether and how the scientific research reaches a goal and whether and how it will be possible to agree upon what is to be considered as the ethical good in a specific situation and how one should behave at best in the flow of everyday life. In these cases, the value of the interaction is not bound to an external goal. Karl Jaspers called this type of interaction "existential communication", where one feels connected to others working towards an abstract overall goal, but concretely struggles with different convictions. Convictions regarding moral good are at stake here, which must be assumed to exist in instrumental actions (Jaspers 1973, pp. 60–73).

According to this dichotomy, actions aiming to take account of epigenetic findings are instrumental actions. Within the framework of epigenetics, influencing of the molecular mechanism remains the goal of the action. Even if epigenetic structures are not directly modified on a molecular level, but rather through a change in behaviour and a change in environmental factors, the justification and the criteria for success of these interventions on the level of organism-environment interactions lie in the respective change in the epigenetic-molecular pattern. Human actions that change oneself and one's environment therefore remain instrumentally aimed towards an external purpose for the action. The actions are a means to reach this specific purpose and gain their value from this relationship. This shows, for example, that if it is observed that direct interventions in the epigenome lead to the desired changes and purposes more efficiently than the interaction on the lifeworld level, the interaction is discontinued, and the direct influencing of the epigenome should be selected as a means instead. Described in this way, the handling of epigenetic findings is characterised by a distinctive tension. In contrast to genetic findings, epigenetic findings open up options for action. However, these options for action remain instrumentally related to the targeted change of epigenetic structures. In other words, they constitute an instrumental relation with oneself and the world. The aspect of action in which the individual is continuously seeking ethically good ways for his own life from his limited perspective, and striving for the non-instrumental good, in contrast, remains unconsidered. However, this is the aspect in which ultimately every form of instrumental action must also be embedded, if this action is to be considered as good, and as sensible and helpful for one's own life praxis in accordance with the corresponding convictions. As said in the words of Jürgen Habermas: In epigenetics, the respective individual faces the problem of "how technically usable findings can be implemented in the practical consciousness of a social lifeworld" in a particularly obvious manner (Habermas 1969, p. 107). Which challenges are then to be faced if epigenetic-instrumental actions should be embedded in social-lifeworld actions can be explained using several conceptual pairs used by Karl Jaspers at central points of his existential philosophy based on Kierkegaard (Jaspers 1973).

As a matter of principle, these conceptual pairs are oriented towards the comparison of an observing and an involved attitude towards knowledge. Knowledge ideals for observing cognition are completeness and coherence of the findings about a specific scope of reality. The cognizer (German "der Erkennende", the man of knowledge) faces reality independently and neutrally. In contrast, the knowledge ideal of integrated cognition is the capacity to be able to behave responsibly oriented towards a concept of good. Thus, this dichotomy includes the above-mentioned distinction between instrumentally usable knowledge and knowledge in the context of a social lifeworld. In the following, the conceptual pairs perception and responsibility, distance and meaning, interaction and communication, and quantity and rank derived by Jaspers from this dichotomy are to be considered in more detail in terms of the problem of the integration of epigenetic knowledge in aspects of life.

5.1 Perception and Responsibility

The first step from the genetic-deterministic world into the world of actions is made by epigenetics itself. While with diseases that are declared to be genetic, the individual can only perceive the genetic prediction without reacting to what will fatefully happen, with a corresponding epigenetic-predictive diagnosis, he can potentially assume responsibility and become an actor. Within the scope of epigenetics, a passive perception of a law, even a statistical law, can become responsible decision and action.

5.2 Distance and Meaning

In the scope of action, however, the epigenetic knowledge initially remains the basis for a purely instrumental behaviour that is instrumentally geared towards the goal of changing the epigenetic structures. This instrumentality of action can also be described as a distance from the actor to the knowledge available to him. Epigenetic-instrumental action requires that the actor, even if he is simultaneously the affected party, understands from an uninvolved expert perspective that epigenetic changes exist and which measures can be taken to reverse this change. In this respect, the individual, even if he is the patient, assumes the perspective of a medical expert in the instrumental action. Furthermore, the individual must now also include his own interests and convictions in the deliberation of which action is instrumentally sensible, because only these can indicate which purpose should then be achieved with the instrumental action. The goal of remaining healthy, which was always assumed to be implicitly matter of course within the scope of epigenetic interventions until now, thus becomes the focus of attention at this point for the first time. In instrumental terms, the individual will—or maybe will not—find this goal in himself, just as he will or will not find the fitting means for this goal in the world. As an instrumental actor, he has the same expert distance from his interests and convictions as from the means of intervention.

At this point, however, the question will inevitably arise for each actor, whether the goal of remaining healthy has the same significance for the actor himself as that which is implicitly granted within the scope of the instrumental-epigenetic action. This question is the question of the individual meaning that disease and health have for the individual person. Depending on the answer to this question, the stress involved for the individual to live a lifestyle that is geared towards epigenetic risk minimisation will be evaluated differently, and to which degree one is willing to accept such a direction in their own life.

5.3   Defined Interactions and Historical Communication

The criteria according to which certain goals can be granted individual meaning cannot be universally defined. In any case, however, meaning has an effect within the conceptions of one's own identity. Meaning is found in those parts of life stories that explain who one is and why one has become that way. Such stories are variable and alterable, because they are based on communication. They express where and how one has searched for the good. This search relies on the opinions and lifestyles of others, the contemplation of which can help one to develop one's own answers. What one believes to have found in the process and what was therefore decisive for individual life phases, can and must not be the last word on this matter from the perspective of the story teller.

Such communicative stories differ from the confirmation of regular interactions. However, in epigenetic-instrumental terms, changes are the result of precisely such regulated interactions. This is mainly true for the molecular explanation of the development of diseases. Nonetheless, if one were looking for the origins of individual convictions regarding the meaning of disease, then any epigenetic explanation would have to postulate some kind of regulated interaction between epigenetic phenomena and the development of these convictions. In order to embed epigenetic knowledge socially and lifeworldly, one must let go of this concept of regulated interaction and look for meaning of superordinate ideals for one's own life within stories about one's own identity that are based on communication.

5.4   Quantity and Rank

Finally, the question about the individual meaning of disease and health causes one to let go of the concept of being able to make decisions relating to the meaning-fulness of changes in behaviour that aim towards epigenetic modification and epigenetic risk minimisation based on quantities. In epigenetic-instrumental terms, this concept is self-evident, because the indication of the probability of occurrence of a disease is the only parameter within the epigenetic analysis that can be put in relation to the urgency of action. One would assume that the higher the probability for the occurrence of a disease, the more urgent it is to counteract. However, if the decision for epigenetic-instrumental action is embedded in stories of the individual meaning of health and disease, the commitment to such a decision does not depend on the degree of the probability of occurrence, but rather primarily on the priority or the rank of the goal of remaining healthy relative to other important life goals. Here, the challenge consists of not succumbing to the obvious pull of the number from an epigenetic-instrumental perspective, but rather to ask oneself about the individual meaning of health within one's own life and the understanding of one's own identity.

6   Conclusion

Epigenetics does not cause a paradigm shift away from a gene-centric and towards an understanding of human development based on environmental interactions. The inclusion of environmental factors in the explanation of genetic activity rather more likely suggests directing action instrumentally towards the effects that this interactive action has on one's own epigenome. It is one of the central ethical challenges involved with epigenetics, to call attention to the limitations of this instrumental logic of action. Ultimately, this task is also a task for the praxis. Genetic consulting that includes epigenetics can only be a good consultation if it is aware of and does justice to the challenges of embedding action instrumentally based on epigenetic structural changes, in social lifeworlds.

References

Boldt, J. (2013). Life as a technological product: Philosophical and ethical aspects of synthetic biology. *Biological Theory, 8*(4), 391–401.

Canning, C. (2008). Epigenetics: An emerging challenge to genetic determinism in studies of mental health and illness. *Social Alternatives, 27*(4), 14–21.

Cremer, T. (2010). Von der Genetik zur Epigenetik und Epigenomforschung - Essay zur Geschichte der Vererbungsforschung und zur Zukunft der prädiktiven Medizin. *Nove Acta Leopoldina, 98*(361), 87–165.

Habermas, J. (1969). *Technik und Wissenschaft als Ideologie.* Frankfurt am Main: Suhrkamp.

Jaspers, K. (1973). *Philosophie II. Existenzerhellung.* Berlin: Springer.

Mansuy, I. (2014). Keine Laune des Schicksals. Interview. Die Zeit vom 27.05.2014 http:// www.zeit.de/2014/22/isabelle-mansuy-epigenetik-hirnforschung, Last access March 2016.

Nicolosi, G., & Ruivenkamp, G. (2012). The epigenetic turn. Some notes about the epistemological change of perspective in biosciences. *Medicine, Health Care and Philosophy, 15*(3), 309–319.

Rakyan, V. K., Down, T. A., Balding, D. J., & Beck, S. (2011). Epigenome-wide association studies for common human diseases. *Nature Reviews Genetics, 12*(8), 529–541.

Rivera, C. M., & Ren, B. (2013). Mapping Human Epigenomes. *Cell, 155*(1), 39–55.

Süring, K. (2010). *Epigenetik – Das molekulare Gedächtnis für Umwelteinflüsse? Hintergrund.* Dessau-Roßlau: Umweltbundesamt http://www.umweltbundesamt.de/sites/default/files/medien/pdfs/epigenetik.pdf, Last access March 2016.

Youngson, N. A., & Whitelaw, E. (2008). Transgenerational epigenetic effects. *Annual Review of Genomics and Human Genetics, 9,* 233–257.

Author Biography

Joachim Boldt is Associate Professor at the Department of Medical Ethics and the History of Medicine, University of Freiburg. Two of his most recent publications deal with synthetic biology and existential ethics, respectively (Boldt J. (ed.) (2016): Synthetic biology. Metaphors, Worldviews, Ethics, and Law. Springer, Berlin) and human vulnerability (Boldt J. (2015): Vulnerabilität, Existenz und Ethik. In Müller O., Maio G. (ed.): Orientierung am Menschen. Anthropologische Konzeptionen und normative Perspektiven. Wallstein, Göttingen). His main areas of research include philosophical and ethical aspects of biotechnologies and conceptual frames and normative justifications of medical interventions into the human body.

Contact: Institut für Ethik und Geschichte der Medizin, Albert-Ludwigs-Universität Freiburg, Stefan-Meier-Str. 26, 79104 Freiburg.

Epigenetics: Biological, Medical, Social, and Ethical Challenges

Kirsten Brukamp

Abstract

Epigenetics, a relatively young discipline in science, is expected to grow steadily and add a fund of knowledge to the life sciences in the future. Regarding the fields of biology and medicine, theoretical questions about epigenetics concern its status in the nexus between nucleotide sequence and gene regulation. Uncertainties for the social sciences relate to the absence of knowledge and the absence of abilities as sociological categories. Ethical problems particularly pertain to the topics of communication about scientific knowledge and about risk as well as to intergenerational justice, reproduction, and the responsibilities of governments and administrations in society. At the moment, a focus of ethical reflection is laid on the education of the public. Information in the media should be adapted to the knowledge that has already been achieved in the sciences. For this purpose, recommendations by learned societies and scholarly organizations can provide standards for orientation.

K. Brukamp (✉)
Duale Hochschule Baden-Württemberg DHBW, Sozialwesen/Gesundheit,
Wilhelmstraße 10, 89518 Heidenheim, Germany
e-mail: brukamp@dhbw-heidenheim.de

K. Brukamp
Universität Rostock, Geschichte der Medizin, Doberaner Straße 140, 18057 Rostock,
Germany

© Springer Fachmedien Wiesbaden GmbH 2017   83
R. Heil et al. (eds.), *Epigenetics*, Technikzukünfte, Wissenschaft und
Gesellschaft / Futures of Technology, Science and Society,
DOI 10.1007/978-3-658-14460-9_7

1   Epigenetics: Still a Young Discipline

Epigenetics is a specialized discipline of biology, in particular of molecular biology. It examines those types of inheritance between cells and organisms that exist independently of the deoxyribonucleic acid (DNA) code, which contains information for protein structures (Eccleston et al. 2007). The attitudes towards epigenetics in society are ambivalent. Is epigenetics more important than genetics, or is its role completely overestimated? Does it imply an aggravation of genetic determinism or liberation from it (cf. Schuol in this volume)? Can epigenetic processes destroy human health even before birth, or are all mechanisms reversible? On the one hand, epigenetic changes appear to possess long-term effects, but on the other hand, the details of modulations are not well known to date.

Epigenetics adds a higher degree of complexity to the established biological disciplines of genetics and genomics—and the same applies to their ethical implications (Knoppers 2009). Because of its relatively recent emergence as a scientific field, epigenetics leads to theoretical difficulties in philosophy in general, insofar as its novel concepts still need to be categorized and understood in depth. Moreover, it is associated with problems in practical philosophy with regard to biological, medical, social, and ethical aspects. Therefore, discourses will arise with regard to theoretical questions about biomedical foundations, topics from the social sciences, and ethical problems of epigenetics.

2   Gene Regulation and Reversibility

Epigenetics may be defined as the science of those heritable cellular characteristics that do not follow directly from the DNA sequence. The exact terminology still remains a matter of debate even among experts in the field: "There has always been a place in biology for words that have different meanings for different people. Epigenetics is an extreme case, because it has several meanings with independent roots." (Bird 2007, p. 396). Examples for epigenetic mechanisms of inheritance are modifications (e.g., methylations) of DNA and histones (Jirtle and Skinner 2007; Bird 2007; Dolinoy 2008; Portela and Esteller 2010). Different epigenetic modifications occur in different tissues and cell types (Rodenhiser and Mann 2006; Jirtle and Skinner 2007).

Philosophical reflection on epigenetics could be named the "meta-physics of epi-genetics". Such deliberation should clarify what the novel properties of epigenetics exactly are. It is not the subject of a new invention, but of a discovery.

Epigenetic mechanisms have been active in living organisms for a long phylogenetic period of time. The fact that environmental factors may exert long-term biological effects on individual organisms is also familiar, such as the impact of nutrition or medication (Jirtle and Skinner 2007; Dolinoy 2008). In science, the basic reversibility of epigenetic modifications has already been investigated. Admittedly, the mechanisms of epigenetics need to be studied further. Academic research has long neglected heritability patterns aside from those stemming from DNA. New insights concern, among other items, effects that reveal themselves over many generations (ibid.). The irreversibility of effects can frequently be observed specifically in developmental biology.

In order to define the role of epigenetics in detail, its main biological concept could be identified as *gene regulation*. According to an old and outdated model, epigenetics surpasses the familiar polarization between gene and gene expression, and it virtually constitutes a novel, third level of reference in molecular biology. In the framework of a new model, however, emphasis could rather be placed on gene functions, transcriptional products (Fraser and Bickmore 2007), bodily plasticities, and environmental interactions (Nicolosi and Ruivenkamp 2012). Then, the gene itself would just be one component necessary to reach a functional goal. Gene regulation or modulation to steer transcriptional activity comprises, at the same time, transcription factor systems, mechanisms of epigenetics, and modifications after transcription.

The DNA sequence of a gene is invariant, and only gene therapy is able to purposefully modify it from externally. The products of gene expression, in contrast, are variable at the levels of messenger ribonucleic acid (mRNA) and protein. The variability of gene expression products is a reaction to nutrition, stress, medication, illegal drugs, doping, and other environmental factors. The gene products may be targeted by controlled pharmacological interventions in vivo. Corresponding to this perspective, epigenetic mechanisms can also be assigned to the side of gene expression.

Reversible *versus* irreversible effects on genes and gene expression need to be distinguished from temporary *versus* permanent effects on phenotypes in order to gain a better understanding of the possible effects of interventions, especially pharmacological interventions. This differentiation takes into account the interactions of mechanisms in molecular biology and in developmental biology. An epigenetic impact may reveal itself after a long time, and it is dependent on the stage of an organism's development. Consequently, there are "long-term effects on development" (Feil and Fraga 2012, p. 98) and "developmental-stage dependent effects" (ibid., p. 99).

A *reversible (epi)genetic effect* may be caused, for example, by a modulation in gene expression after the use of medication. The result can be a *temporary* or a *permanent phenotypic effect* (Feinberg 2007). An infamous example for a permanent phenotypic effect is the grave leverage on developing embryos when pregnant women take thalidomide (which is better known to the German public under the brand name Contergan®) during the first three months of pregnancy (Ito et al. 2011). An *irreversible (epi)genetic effect* is caused, among others factors, by gene therapy. From this, a *permanent phenotypic effect* may follow for the germ line, such as after certain types of gene therapy (Kay 2011). As a result, distinctions and considerations should be made, at all times, between genetic and epigenetic levels, reversible and irreversible molecular processes, and temporary and permanent phenotypic consequences in order to promote transparency in the discussions about (epi)genetic problems.

3 Knowledge, Ignorance, and Science Communication

The fund of knowledge in epigenetics is currently still rapidly expanding. A fruitful field is thereby being opened for studies in the sociology of knowledge. In the sociology of knowledge, the *absence of knowledge,* also termed *non-knowledge, lack of knowledge,* or *ignorance,* is not only distinguished from *knowledge,* but different kinds of ignorance may be differentiated as well (Wehling 2006; Brukamp 2013). A crucial distinction pertains to *known, specified ignorance* (Merton 1968), i.e., ignorance of which people are aware, and *unrecognized, unknown ignorance* (Kerwin 1993), which is not consciously accessible. Missing knowledge in specified ignorance can in turn be divided into *knowledge that has not been reached* as of yet, *knowledge that cannot be reached* in principle, and *knowledge that is not actively being sought after* and is thus unwanted (Brukamp 2013). Knowledge that has not been reached to date includes, for example, the details of many epigenetic mechanisms. Knowledge that cannot be reached contains causal relationships in the past. Knowledge that is not actively being sought after pertains to the results of morally objectionable experiments, which are not being carried out. Unknown ignorance prevailed in the past when both the existence and the relevance of epigenetics had not yet been recognized. The concept for this classification of knowledge and ignorance can be applied to *abilities* and *inabilities* as well.

Epigenetic knowledge that has not been reached as of yet concerns scientific analysis and medical diagnostics. For scientific interventions and medical therapies, in turn, deficits persist regarding options for epigenetic modifications and modulations, even though some therapies have already been approved for clinical practice (Tycko et al. 2013). The problematic constellation can be summarized as follows: Epigenetic mechanisms and effects have not yet been investigated in sufficient detail, a fact that points to the knowledge that has not yet been reached. Due to the absence of technical abilities that have not been acquired yet, research regarding therapeutic interventions is still deficient, despite the hypothesis that all epigenetic modifications are reversible. An ability that cannot be acquired in principle is the assurance that one can causally influence later effects uncompromisingly. Permanent and irreversible phenotypic effects occur at times despite resulting from merely temporary changes during early stages of human development. The subsequent treatment can therefore only be oriented towards providing symptomatic treatment and relief.

From the perspective of the sociology of knowledge, ethical problems arise for the areas of science communication and the response to information deficits. At the moment, so-called *genetic determinism* is still a matter of wide debate in the public, in contrast to the field of biology (cf. Schuol in this volume). Its basic assumptions are that the human constitution is largely determined by genes and that education, nutrition, and environmental influences, in contrast, play a relatively small role. This attitude, however, no longer corresponds to the newest scientific results (Bird 2007). For instance, genetically identical mice with the same DNA endowments can acquire completely different features for body weight, coat color, and morbidity, depending on which diet they consume and which environmental influences act on them (Jirtle and Skinner 2007; Dolinoy 2008). Thus, a new, comprehensive model needs to be developed that takes the multifactorial influences on organisms into consideration. The question remains as to when enough research data will be available to appropriately accomplish this task. Afterwards, the model needs to be communicated to the interested public parties, and this process may prove a hurdle, too. Public dismissal of genetic determinism could potentially draw attention to the social determinants of health (Loi et al. 2013).

The general public suffers from deficits with regard to information on the latest insights into cell, molecular, and developmental biology that refers to epigenetic influence on human attributes. Building on this new knowledge, which experts have already recognized, strategies need to be developed as to how to inform and educate the public about the possible outcomes of epigenetic inheritance. For example, health care professionals could become active in counseling for health promotion and disease prevention on the basis of the novel medical or political guidelines that are expected to come. Since biological facts are sometimes highly complex, they pose a challenge to determining the adequate way to provide education about biology and medicine so that humans are consequently empowered to arrive at their own decisions about their lifestyles.

In summary, with regard to the topics of *science communication* and *the response to information deficits*, new educational strategies need to be devised concerning how to supply the public with genetic and epigenetic information. Remaining questions relate to the methods to convey knowledge about molecular processes and the means to tackle beliefs about the existence of genetic determinism. Not only genes are determinants of human lives, but so also are nutrition, medication, education, and the environment in general (Jirtle and Skinner 2007; Dupras et al. 2012). The public discourse needs to begin at the right moment to address some of the uncertainties that are still present within biomedical research (cf. Seitz in this volume). In order to arrive at appropriate recommendations in due time, sufficient data need to be taken into account from the results of biological research and from the outcomes of medical therapy.

4   Intergenerational Justice for Reproduction

A topic that plays an important role in the assessment of epigenetics is *intergenerational justice for reproduction* (Rothstein et al. 2009, Rothstein 2013; Guibet Lafaye 2013). As a general rule, biological research projects uncover mechanisms to gain knowledge that often cannot be transformed into reasonable options for medical interventions until much later. Corresponding to this frequently observed phenomenon, a convincing prognosis for future projects is that more epigenetic mechanisms will become known at every moment, whereas possible interventions and therapies will frequently not be available for some additional period of time. Humans acting autonomously decide for themselves how to use the accessible information, for instance for designing their nutrition plan. Vulnerable groups, on the other hand, include particularly vulnerable humans with physical or social restrictions, and they are especially in the focus of ethics because of the principle of beneficence (cf. Beauchamp and Childress 2013).

Two special features of epigenetic mechanisms are the facts that they are already active during development in embryos, fetuses, and children and that some of their consequences can be fully foreseen only much later (Feil and Fraga 2012). Familiar *models*, which have already been examined in the past, can help to better evaluate and judge the specific problems of epigenetic effects during development. Based on previous problems in medicine and topics in medical ethics, models for epigenetics may concern how a pregnant woman deals with her body. Candidates for such models are diet and the use of legal or illegal drugs (e.g., alcohol) during pregnancy.

The *fetal alcohol syndrome* (FAS) (Feldmann et al. 2007; Landgraf and Heinen 2012; Landgraf et al. 2013) is regarded as a subgroup of the *fetal alcohol spectrum disorder* (FASD). Formerly known by its synonym alcohol embryopathy, fetal alcohol syndrome is characterized by the effects of alcohol on embryos and fetuses. A partial phenotypic effect is possible as well. The etiology, i.e., the origin, of fetal alcohol syndrome is alcohol use during pregnancy. The pathogenesis, i.e., the disease process, involves damaging consequences on developing tissues and organs, among others on the brain. The children who are affected may stem from women with a diagnosis of alcohol use disorder, but also from women who rarely use alcohol during pregnancy. The children are infrequently diagnosed, probably due to a lack of awareness and because of social tabooization.

The somatic *symptoms* of fetal alcohol syndrome (Feldmann et al. 2007; Landgraf and Heinen 2012; Landgraf et al. 2013) include short stature, under-weight, developmental retardation, damage to internal organs, congenital heart defects, damage to the heart muscle, decreased muscle tone, small distal phalanges of the fingers, and typical changes of the features of the head, like small eyes with drooping eyelids (ptosis), oblique palpebral fissures, ear deformities, cleft forma-tions, flat upper jaws, thin upper lips, a smooth philtrum between the nose and upper lip, and small teeth with increased distances. Neurological and psychiatric symptoms are developmental retardation, cognitive deficits, poor concentration, learning disabilities, hyperactivity, impaired social behavior, aggressiveness, and antisocial behavior. The fetal alcohol syndrome is viewed as the most common non-genetic cause of mental retardation and as avoidable.

The example of fetal alcohol syndrome illustrates a conflict between different principles in medical ethics (cf. Beauchamp and Childress 2013), namely between the respect for the autonomy of people and the prevention of harm to others. This conflict can presumably also be identified for epigenetic mechanisms that have not yet been comprehensively detected. On the one hand, a woman's *autonomy and self-determination* regarding her pregnant body should be respected. On the other hand, the *welfare* of the developing child should be protected by preventing harm that affects it. The legal regulations in Germany state that the government is not authorized to order direct bodily interventions or their omissions for citizens under normal circumstances. Here, parallels can be identified with regard to the ethical, legal, and political principles for fetal alcohol syndrome on the one hand and adult epigenetic diseases, which will be diagnosed in the future, on the other hand. Insofar, the public handling of fetal alcohol syndrome can be regarded as a model that can also be applied to epigenetically determined diseases.

Consequently, obvious measures comprise providing information and education to a pregnant woman and the later parents, appeals to the pregnant woman to care for her developing child, and recommendations by independent academic organizations and learned societies in medicine. With regard to patient and client education in general and the *counseling* of pregnant women in particular, clear and understandable information is necessary, and it should be as practical and accessible as possible. Obstetricians and midwives, for example, could distribute it during personal contacts, as could modern media *via* rather impersonal routes. Special attention should be devoted to the communication of risks, particularly the representation of probabilities. Risks, for instance, can be displayed as absolute, relative, or dependent on an individual change after an intervention. A rough classification as normal, high, or low could be appropriate, and numbers, words, or visual presentations might be chosen.

5 Responsibility in Society with Regard to Reproduction

The term "responsibility" has been identified as essential for the ethics of epigenetics (Hedlund 2010; Chadwick and O'Connor 2013; cf. Boldt and Schuol in this volume as well). However, no immediate, tangible instructions for action follow from this insight. Reflecting on the significance of epigenetics in reproduction leads to the *problem of the roles to be played by governments and administrations.* Up to now, it has remained uncertain whether the new scientific results imply that the government must become more active (cf. Robienski in this volume) in regulating food or recommending its use. A pregnant woman's diet exerts a special influence on a developing child, as the example of fetal alcohol syndrome illustrates. Consequently, the question follows whether pregnant and breast-feeding women should be allowed to continue eating whatever they want in accordance with ethical arguments in favor of human autonomy and self-determination. In Germany, a foundation for disabled children initiated advertisements in the public sphere in order to discourage women from drinking alcohol during pregnancy. The German Federal Center for Health Education (i.e., "Bundeszentrale für gesundheitliche Aufklärung", BZgA, in German) offers brochures on this topic.

Overall, it is questionable whether administrative sanctions in the negative sense may be instituted with regard to a *fundamental need like eating*, unless arguments stemming from damages can unambiguously be traced back to the food that pregnant women take in and its epigenetic effects. Dietary choice and intake belong to the essential capabilities of every living organism, and they constitute a fundamental act of self-determination. Related interventions by administrative authorities at the level of the individual would have to be justified extraordinarily well. For this, an ethically important aspect would be the potential harm to others. The detectability of a pregnancy poses a practical problem for education, which should be politically legitimized and tailored towards the individual situation. As long as a pregnant woman does not seek attention and care in the medical system, she cannot be informed by physicians and midwives about any potential risks for her target group. Thus, the most appropriate way currently is to inform the public, for example by means of information campaigns. Recommendations for a woman's diet before and during pregnancy already exist (Koletzko 2013). Aside from this, reliable new results of scientific research are presently not available and thus cannot serve to justify further policies.

A problem related to nutrition during pregnancy is that concerning *functional food* (i.e., nutraceuticals). So far, the medical sequelae—both the advantages and the disadvantages—remain unclear for the development of and the trade with dietary supplements and functional food products that contain virtually medicinal ingredients. Gender-specific differences of acceptance and use behavior will most likely occur, and different social groups with distinct motives for use can be identified (Bundesinstitut für Risikobewertung et al. 2013). Parallel assumptions follow for differences in the acceptance, use, and motivation between groups of pregnant women, when an analogy, as a means for comparison, is drawn between problems with functional food and nutrition in pregnancy. Consequently, more detailed sociological investigations should be conducted concerning these phenomena, for example with the intention of providing information in a manner that is appropriate for a target group and influences its behavior.

In summary, epigenetics requires *responsibility in society*, and the *tasks of governments and administrations* need to be identified. Both ignorance and inability continue to persist in biomedicine and take the form of the absence of knowledge and abilities that have not yet been reached. For this reason, the first conclusions would be to pursue additional research in the natural sciences and to begin a public discourse without scandalization (cf. Seitz in this volume). So far, both governments and health care systems have demonstrated their responsibility for providing support to pregnant women by issuing recommendations for nutrition and for the prevention of toxic substances. Correspondingly, counseling for phenotypically relevant epigenetic effects needs to be optimized. Additional regulations can, under the current conditions of ignorance, probably not be justified when they relate to situations in which epigenetic effects occur without any obvious toxic impact.

6   Epi-Genetic Meta-Insights?

The present status of scientific research in epigenetics does not allow a comprehensive assessment of the future significance of epigenetics for the natural sciences, the social sciences, the humanities, and ethics. The insight that nutritional components and environmental factors can exert medium- and long-term influences on living organisms is not a new finding. In contrast, novel data concern the mechanisms that are investigated in epigenetics as a field of the life sciences.

In summary, epigenetics appears as a discipline that is still at the beginning of a highly promising development that permits progress to be made in biology and medicine. Partly owing to the aspects of ignorance that are currently prominent, epigenetics as a science encounters theoretical and practical as well as social and ethical problems that cannot be fully evaluated at the moment (Pickersgill et al. 2013). These problems have not been amenable to easy solutions thus far.

References

Beauchamp, T. L., & Childress, J. F. (2013). *Principles of biomedical ethics*. Oxford: Oxford University Press.

Bird, A. (2007). Perceptions of epigenetics. *Nature, 447*, 396–398.

Brukamp, K. (2013). Nichtwissen in der Neuromedizin: Wissenschaftliches Wissen und Nichtwissen bei gegenwärtigen Neurointerventionen im Gehirn. In C. Peter, & D. Funcke (Eds.), *Wissen an der Grenze. Zum Umgang mit Ungewissheit und Unsicherheit in der modernen Medizin* (pp. 309–338). Frankfurt am Main: Campus.

Bundesinstitut für Risikobewertung (Eds.), Rehaag, R., Tils, G., Röder, B., Ulbig, E., Kurzenhäuser-Carstens, S., Lohmann, M., Böl, G.-F. (2013). *Zielgruppengerechte Risikokommunikation zum Thema Nahrungsergänzungsmittel. Abschlussbericht.* Berlin: BfR-Wissenschaft 03/2013. www.bfr.bund.de/cm/350/zielgruppengerechte-risikokommunikation-zum-thema-nahrungsergaenzungsmittel.pdf. Last access March 2016.

Chadwick, R., & O'Connor, A. (2013). Epigenetics and personalized medicine: Prospects and ethical issues. *Personalized Medicine, 10*(5), 463–471.

Dolinoy, D. C. (2008). The agouti mouse model: An epigenetic biosensor for nutritional and environmental alterations on the fetal epigenome. *Nutrition Reviews, 66*(Suppl 1), 7–11.

Dupras, C., Ravitsky, V., & Williams-Jones, B. (2012). Epigenetics and the environment in bioethics. *Bioethics, 28*(7), 327–334.

Eccleston, A., DeWitt, N., Gunter, C., Marte, B., & Nath, D. (2007). Epigenetics. *Nature, 447*, 395.

Feil, R., & Fraga, M. F. (2012). Epigenetics and the environment: Emerging patterns and implications. *Nature Reviews Genetics, 13*, 97–109.

Feinberg, A. P. (2007). Phenotypic plasticity and the epigenetics of human disease. *Nature, 447*, 433–440.

Feldmann, R., Löser, H., & Weglage, J. (2007). Fetales Alkoholsyndrom (FAS). *Monatsschrift Kinderheilkunde, 9*, 853–865.

Fraser, P., & Bickmore, W. (2007). Nuclear organization of the genome and the potential for gene regulation. *Nature, 447*, 413–417.

Guibet Lafaye, C. (2013). Ethical issues raised by research in epigenetics. Presentation at the UNESCO Chair in Bioethics 9th World Conference (November 19–21, 2013). www.academia.edu/5161957/Ethical_issues_raised_by_research_in_epigenetics. Last access March 2016.

Hedlund, M. (2010). Epigenetic responsibility. Papper att presenteras på Statsvetenskapliga förbundets årsmöte Göteborg 30 september–2 oktober 2010. Statsvetenskapliga institutionen, Lunds universitet. pol.gu.se/digitalAssets/1315/1315810_maria-hedlund-epigenetic-responsibility.pdf. Last access March 2016.

Ito, T., Ando, H., & Handa, H. (2011). Teratogenic effects of thalidomide: Molecular mechanisms. *Cellular and Molecular Life Sciences, 68*, 1569–1579.

Jirtle, R. L., & Skinner, M. K. (2007). Environmental epigenomics and disease susceptibility. *Nature Reviews Genetics, 8*, 253–262.

Kay, M. A. (2011). State-of-the-art gene-based therapies: The road ahead. *Nature Reviews Genetics, 12*, 316–328.

Kerwin, A. (1993). None too solid: Medical ignorance. *Science Communication, 15*, 166–185.

Knoppers, B. M. (2009). Genomics and policymaking: From static models to complex systems? *Human Genetics, 125*(4), 375–379.

Koletzko, B. (2013). Ernährung in der Schwangerschaft. Für das Leben des Kindes prägend. *Deutsches Ärzteblatt, 110*(13), A 612–A 613.

Landgraf, M., & Heinen, F. (2012). *S3-Leitlinie Diagnostik des Fetalen Alkoholsyndroms. Kurzfassung.* AWMF-Registernr.: 022–025. www.awmf.org/uploads/tx_szleitlinien/022-025k_s3_Fetales_Alkohol-Syndrom_Diagnostik_Kurzfassung_2012-12.pdf. Last access March 2016.

Landgraf, M. N., Nothacker, M., Kopp, I. B., & Heinen, F. (2013). Diagnose des Fetalen Alkoholsyndroms. *Deutsches Ärzteblatt, 110*(42), 703–710.

Loi, M., Del Savio, L., & Stupka, E. (2013). Social epigenetics and equality of opportunity. *Public Health Ethics, 6*(2), 142–153.

Merton, R. K. (1968). *Social theory and social structure.* New York: Free Press.

Nicolosi, G., & Ruivenkamp, G. (2012). The epigenetic turn. Some notes about the epistemological change of perspective in biosciences. *Medicine, Health Care, and Philosophy, 15*(3), 309–319.

Pickersgill, M., Niewöhner, J., Müller, R., Martin, P., & Cunningham-Burley, S. (2013). Mapping the new molecular landscape: Social dimensions of epigenetics. *New Genetics and Society, 32*(4), 429–447.

Portela, A., & Esteller, M. (2010). Epigenetic modifications and human disease. *Nature Biotechnology, 28*(10), 1057–1068.

Rodenhiser, D., & Mann, M. (2006). Epigenetics and human disease: Translating basic biology into clinical applications. *CMAJ, 174*(3), 341–348.

Rothstein, M. A. (2013). Legal and ethical implications of epigenetics. In R. L. Jirtle, & F. L. Tyson, *Environmental epigenomics in health and disease. Epigenetics and complex diseases* (pp. 297–308). Berlin/Heidelberg: Springer.

Rothstein, M. A., Cai, Y., & Marchant, G. E. (2009). The ghost in our genes: Legal and ethical implications of epigenetics. *Health Matrix Clevel, 19*(1), 1–62.

Tycko, J., Fields, D., Cabrera, D., Charawi, M., & Kaptur, B. (2013). The potential of epigenetic therapy and the need for elucidation of risks. *Penn Bioethics Journal* VII (ii): 1–4. Philadelphia: University of Pennsylvania. 2013 http://igem.org/wiki/images/a/ae/Penn_iGEM_2013_PBJ_Article.pdf. Last access March 2016.

Wehling, P. (2006). *Im Schatten des Wissens? Perspektiven der Soziologie des Nichtwissens.* Konstanz: UVK.

Author Biography

Kirsten Brukamp is the Professor of Health Sciences at the DHBW Cooperative State University and a lecturer in history, theory, and ethics of medicine at Rostock University in Germany. She completed studies in medicine, philosophy, and cognitive science, and she has conducted several research projects in the fields of cell and molecular biology. Her publications (selection) include "Hypoxia and podocyte-specific Vhlh deletion confer risk of glomerular disease" (Am J Physiol Renal Physiol 2007; with co-authors); "Transplant nephrectomy: histologic findings—a single center study" (Am J Nephrol 2014; with co-authors); "Right (to a) diagnosis? Establishing correct diagnoses in chronic disorders of consciousness" (Neuroethics-Neth 2013). Research interests: life sciences, ethics of medicine and biology, neurophilosophy.

Contact: Duale Hochschule Baden-Württemberg DHBW, Sozialwesen/Gesundheit, Wilhelmstraße 10, 89518 Heidenheim, Deutschland; Universität Rostock, Geschichte der Medizin, Doberaner Straße 140, 18057 Rostock, Deutschland.

Learning from and Shaping the Public Discourse About Epigenetics

Stefanie B. Seitz

Abstract

Epigenetics is one of a group of newly developing technosciences whose applications are just starting to leave the sphere of pure science. What is currently known about epigenetics is not only the basis for grand visions, but also for first applications especially in the medical sector. Epigenetics is therefore already of relevance to our daily lives. However, the consequences that epigenetics will have for society have hardly been investigated in detail and pose a number of challenges. Scientific and public discourses about epigenetics and its consequences have commenced, and are featuring in the media. This paper illuminates the extent to which technology assessment could (and should) incorporate the public discourse into its deliberations, and whether active steps to shape this discourse could contribute to the elaboration of a responsible approach to a new scientific discipline and its applications that enjoys societal support.

1   Introduction

Even during the 'early years of genetics', researchers observed phenomena that were difficult to explain with the theories of classic genetics expounded by Gregor Mendel and Charles Darwin (Charlesworth and Charlesworth 2009). Today, many

S.B. Seitz (✉)
DBFZ – Deutsches Biomasseforschungszentrum gGmbH,
Torgauer Str. 116, 04347 Leipzig, Germany
e-mail: stefanie.seitz@dbfz.de

© Springer Fachmedien Wiesbaden GmbH 2017
R. Heil et al. (eds.), *Epigenetics*, Technikzukünfte, Wissenschaft und
Gesellschaft / Futures of Technology, Science and Society,
DOI 10.1007/978-3-658-14460-9_8

95

of these phenomena are subsumed under the term 'epigenetics'. This term was coined in 1942 by Conrad Waddington (1905–1975): Waddington described development and inheritance as 'the cross-talk between genetic information and the environment' (cf. Bird 2007, p. 396). Today, 70 years later, epigenetics is a discipline in its own right within the biosciences, and the definition of the term differs sharply from that put forward by Waddington (cf. Schuol in this volume). In the recent literature, the reader often comes across explanations that the term is a combination of 'genetics', i.e. the study of heredity (and genes as hereditary factors), and the Greek prefix 'epi', and epigenetics hence means 'additional to genetics' or 'on top of genetics'. This diverges from Waddington's etymology, but describes the current understanding of epigenetics very well—in particular as it relates to genetics.

Epigenetics currently deals with the mechanisms of gene regulation and heredity[1] independently from the DNA sequence of the genes (cf. Walter and Hümpel in this volume). This is understood to refer principally to three processes (cf. e.g. Youngson and Whitelaw 2008): Firstly, the acquired modification of particular DNA bases (e.g. DNA methylations), secondly, modifications to chromatin (e.g. as a result of changes in histone composition and structure) and, thirdly, RNA-mediated gene regulation mechanisms (e.g. what is known as 'RNA interference'). In this respect, it should be clarified that 'independence from the DNA sequence' merely means that these mechanisms modify the phenotype (the entirety of an organism's observable characteristics, incl. its metabolism) by altering gene expression without the genotype (the entirety of the hereditary information) undergoing any change. Otherwise, it goes without saying that the DNA sequence plays an important role in the identification of the target genes for epigenetic regulation. Apart from this, the components of the epigenetic regulatory systems themselves (proteins and small, non-coding RNA strands) are of course dependent on the DNA sequence and the regulation of their own genes. In addition to this, like all the other components of the cell, they too are integrated into a complex regulatory network in which, ultimately, things never happen 'independently from one another'. It should be noted that there is no generally accepted consensus about which mechanisms can be assigned to epigenetics and which can not. To a very great extent, it is incumbent upon the scientists who publish in the field to allocate

---

[1]The degree to which epigenetic markers are inherited and epigenetic mechanisms can have transgenerational effects is still controversial. Since an epigenetic effector (e.g. an environmental stimulus) can influence up to three generations simultaneously (a pregnant woman, her foetus and the foetus's germ line), it is very difficult to supply scientific proof of such effects.

phenomena to these categories. The imprecision of such definitions is something epigenetics shares with other recent disciplines and emerging technologies (cf. Heil et al. in this volume).

In essence, epigenetics combines findings from various fields of research in the biosciences and has therefore enormously expanded the horizon of understanding in relation to the regulatory mechanisms that influence the observable characteristics of living organisms, as well as developing a new, more extensive understanding of heredity and, in the end, evolutionary theory too. This has prompted a revival of interest in Jean-Baptiste Lamarck, whose transformism[2] had once been consigned to gather dust in the archives of science history as a mistaken theory, as well as speculation informed by the findings of epigenetics that he may possibly have been 'ahead of his time' (cf. e.g. Handel and Ramagopalan 2010). Although it has not so far been possible to settle this question, it is now evident the DNA sequence alone does not constitute the biological blueprint for living organisms, but there is 'something else' that is, above all, still susceptible to 'external' influences. This means what is known as the 'gene dogma' (cf. Schuol in this volume) appears to have been superseded! Some scientists even speak of epigenetics as the 'science of change' (Weinhold 2006), while authors of popular science books have proclaimed the 'ousting of the gene' (e.g. Kegel 2009 and Blech 2010).

What is more: Epigenetics is no longer 'just' a field of research in the biosciences, but can now be classified as one of the *new and emerging sciences and technologies* (NESTs),[3] which are leaving the scientific sphere and posing a series of new challenges to the societies that they penetrate. Hitherto, however, the discourse about these challenges and their handling has to a very great extent been conducted without the participation of an interested public. In the rest of this article, I will explore what added value the involvement of the public has for the academic debate, what scientific questions are thrown up about the conditions under which a public discourse could be conducted, and whether a discourse actively shaped by researchers—in the spirit of Responsible Research and

---

[2]Lamarck believed that living organisms acquired new properties because changed environmental conditions forced them to adopt new habits. These properties were then passed on to the organism's descendants. The classic example was that of giraffes, which had to stretch their necks as acacia leaves grew ever higher and therefore came to have longer necks from generation to generation. It was just this part of his theory that later came to be known as Lamarckism and was regarded for a long time as having been disproved (cf. Lefèvre 2001).

[3]This term was used for the first time in the 6th EU Framework Programme for Research and Technological Development in 2003.

Innovation (RRI)—could contribute to the elaboration of a responsible approach to this scientific discipline and its applications that enjoys societal support.

2 Epigenetics from the Perspective of Technology Assessment

Technology assessment (TA) deals essentially with the intentional and unintentional consequences that technologies and technological developments have for society, and seeks knowledge that may guide the management of socio-technical systems (cf. Grunwald 2010). In the light of the 'control dilemma' (Collingridge 1982), which assumes technologies can be easily shaped in their early phases of development, it is a particular concern for TA to begin their assessment within this time slot (cf. Decker in this volume). This is rewarding for all new, emerging technologies from which solutions to the 'grand challenges' faced by society are hoped, even though there is simultaneously great uncertainty with regard to their consequences (cf. Swierstra and Rip 2007; Torgersen 2013). This makes epigenetics a suitable topic for TA.

Epigenetics can be contextualised as what is known as a 'technoscience', 'in which the traditional boundaries between (cognition-oriented) natural science and (application-oriented) technoscience dissolve, and fundamental scientific research is placed in a technical context of commercial exploitation right from the beginning' (Grunwald 2012, p. 10): After all, epigenetics is no longer just a field of pure research; indeed, its results have already led to a series of applications—and many more are in development. Today, epigenetic approaches belong to the standard repertoire of biomedicine, above all in medical diagnostics and therapy, but the findings of epigenetic research are having an influence on society far beyond this field as well (cf. Heil et al. in this volume). At the same time, the consequences of these findings for our society and individuals' daily lives are still largely unexplored. Up until now, only a few scientific investigations into the legal, ethical and social impacts of epigenetics have appeared—one of the few comprehensive papers has been published by Rothstein et al. (2009). However, the growing number of scientific lectures and congresses held on this topic shows that disciplines outside the biosciences are paying ever more attention to epigenetics. In this respect, though, it has not been clarified where the dividing line between epigenetics and 'conventional' genetics runs (if such delimitation is at all meaningful) or what the true potential of epigenetics could be. The consequent legal, ethical and social implications which, further to the individual's own responsibilities, would seem to require the state to assume a particular duty to protect, have not yet been spelled out either (cf. Heil et al. in this volume). The question of the extent to which the public should be involved in the negotiating process and decisions concerning these questions is the subject of the discussion in the sections below.

3 TA's View of the Public Discourse About Epigenetics

3.1 Definitions and Benchmarks for 'Public Discourse'

To begin with, I would like to clarify what is meant by 'public discourse' in the present article. The term 'public' has always been one 'of remarkable wooliness', as Negt and Kluge (1972, p. 17) noted. Depending on which disciplinary context the term is used in, it has various meanings: While an empirical/analytical approach is generally followed in communications science, normative-functional approaches are found more frequently in the social sciences (cf. Donges and Imhof 2001). The term 'public' is used by communications science as the opposite of the private sphere and private communication. Here, the public is distinguished by the 'non-exclusive character of the group addressed in a communication' (Maletzke 1972, p. 24). According to Gerhards and Neidhardt (1991), three levels are differentiated (political public, media public and spontaneous public) in order to do justice to various communicative situations and assign actors to roles (speaker, mediator or audience) in those situations (cf. Donges and Imhof 2001). In contrast to this, the definitions of these terms used in the social sciences always convey normative ideas about the functions of the public within a democracy as well. Accordingly, such definitions are closely linked with the theory of democracy in which they are rooted. For example, in his 'discourse ethics', an ideal-typical discourse model, Jürgen Habermas defines 'the public' as a network for communication, 'within which the discursive formation of opinions and decision-making by an audience of citizens can take place' (Habermas 1962, p. 42).

In the present article, 'the public' means interested citizens and the groups that represent their interests (e.g. civil society organisations), as is customary in the field of participatory TA (cf. Hennen et al. 2004, p. 15ff.). Consequently, 'public discourse' is understood as a form of discourse—here on epigenetics and the societal dimensions of its applications—that is accessible to interested citizens and the groups that represent their interests. This means arguments advanced by actors in science and politics are made publicly accessible, and the public itself is given the opportunity to introduce its own arguments into the discourse.

If this is considered as given, the mass media come to have an important role. For instance, Lehmkuhl (2011) notes in relation to synthetic biology that, if the opportunities and risks of a new technology are to be addressed by society and therefore become a subject of public discourse, 'the mass media will play a key part. For—at least up until now—the production of the public realm has been and still is largely associated with the mass media. It is only through the media that public attention can be concentrated on a topic, only they are capable of helping competing positions and interpretations to gain societal recognition, and therefore make them points of reference for political decisions' (ibid., p. 2, own translation). A survey of the detectable public discourse on epigenetics will be found elsewhere (Seitz and Schuol in this volume). It is, however, to be noted that hitherto this discourse has been limited to the articulation of the parties' own standpoints through the printed media and also that only actors on the peripheries have been expressing their ideas, i.e. representatives of civil society organisations and scientists. So far, a discourse with a dialogic character has merely been conducted in a few expert panel discussions and some Internet forums.

3.2   Why Is the 'Public Discourse' on Epigenetics Relevant?

The question of why technology assessment should also monitor new, developing technologies is explained in the literature using the 'Collingridge or control dilemma' (cf. Decker in this volume). This says that the prospects of certain knowledge about a technology's consequences become better the more highly developed the technology is, i.e. the more is known about its production conditions, contexts of use and disposal processes. However, the opportunities to shape the technology or influence its impacts become worse at the same time. When a development is well advanced or has already concluded, this makes any change of course difficult if not impossible, nevertheless this particular few is discussed increasingly controversial (cf. Grunwald 2010). With the arrival of 'Responsible Research and Innovation' (RRI; von Schomberg 2013) in the world of TA, the motivation to monitor technical developments at an early stage and steer them where necessary has grown stronger once again. RRI endeavours to direct the entire research and innovation process towards societal needs, thus making it more responsible and sustainable. Stilgoe et al. (2013) define four dimensions of responsibility under RRI that that would, among other things, strengthen the involvement of all affected stakeholders and the general public in the relevant decision-making processes. Applied to epigenetics (contextualised as a NEST), this would mean that, thanks to the early stage of development it is at, political decision-makers, scientists and the interested public would be able to collaborate in (i) arriving at standards for the responsible handling of this scientific discipline and its applications, (ii) identifying risks early on and averting them, and (iii) tackling ethical challenges appropriately.

There are very different opinions within the scientific community concerning the extent to which the public should be involved in decisions about the development of technology and what status their participation should have. The foundations for these ideas are always questions such as: Who initiates public participation? For what purpose? What group of persons will be involved in it? Who ensures the quality of the process? And what happens with the results? How these and similar questions raised in connection with public participation are answered depends very much on the authors' standpoint with regard to the role of the public in the political system (Stirling 2008; Walk 2013). For example, it would be possible to view public participation in political decisions as an integral component of representative democracies, but at the same time to identify precisely this as a problem for the legitimation of participation.

Instructive insights into the arguments for and against public participation, and the historical development of the positions on which they are based are offered by Bauer et al. (2007). The authors posit three paradigms for the 'deficit model' they use to explain the deficits in debates about new technology. These paradigms have dominated the discussion successively, but not necessarily replaced each other: In the first phase ('scientific literacy' or technocracy, as of 1960), political decision-makers believe the public does not know enough about science or has no interest in it and is therefore not qualified to be involved in the decision-making processes on technology policy. As a consequence, these decision-making processes take place without public participation. In the second phase ('public understanding of science', after 1985), political decision-makers assume that attitudes will become more positive as more comes to be known about a technology, and the focus is placed on public education. Finally, in the third phase ('science and society', from 1990 on), it is recognised that there are also deficits among the scientific community, which harbours prejudices against the public, and fails to find suitable modes of communication and mediation. The consequence of this is also a lack of public trust in decisions about technology policy. According to this theory/rhetoric, laypeople's deficits therefore stand alongside the experts' deficits, and deliberative/participatory formats and public discourse should help remedy the situation. Efforts to deal with these deficits by involving the public are often linked rhetorically with the creation of an innovation-friendly climate among the population, which is why this linkage is increasingly to be found in programmatic papers on research and innovation policy (e.g. Horizon 2020 or the High-Tech Strategy 2020 for Germany adopted in 2014), and some scientists are already claiming to see a 'participatory turn' (Jasanoff 2003). Public participation is very clearly being instrumentalised here.

Another approach to the purpose of public participation is pursued by participatory technology assessment. This stream of TA comprehends the participation of the public under participatory and deliberative procedures as an integral aspect of its research: 'Participatory TA engages with ideas concerning new forms of cooperation and dialogue between the scientific community, policymakers and the public, driven by the conviction that the comprehensive evaluation of new technologies relies on incorporating the values by which societal groups are guided and those groups' interests' (Hennen et al. 2004, p. 4). In consequence, participatory TA stands in the tradition of Habermas (1970), who argued that informed public debate led to better policy decisions (cf. Hennen 2012, p. 29f.). It therefore consciously integrates participatory elements into the process of political deliberation within the framework of the mandates for TA activities. Here, the involvement of the public primarily serves the generation of knowledge, e.g. about what hopes and fears the population associate with epigenetics, what perceptions of risk are widespread and what governance measures are desired. In consequence, the normative demands made of public participation also become less onerous because the question of its anchoring in the political system has been clearly answered here, and the expectations with regard to the incorporation of its results into decision-making processes are not so high. To what extent and through what channels the public should be involved depends on the concrete parameters, and adapted participation formats are selected accordingly (e.g. Bütschi and Nentwich 2000; Stirling 2008; Bijker 2013).

As far as this issue is concerned, Bijker (2013) argues that it has to be decided who ideally ought to be involved in political decision-making concerning the handling of a technology's uncertainties on the basis of what is known about it and how far it has developed. If established technologies whose risks are well known throw up concrete questions, it is most expedient to simply obtain advice from experts in order to develop solutions for these concrete risks. When this is done, expert consultations are, above all, a highly suitable way of clarifying the priorities for the studies that are carried out, developing options for the resolution of problems, and evaluating the use of a particular technology in the light of the scientific findings that are available on its risks and opportunities (cf. Hennen et al. 2004). The situation is different if such decisions have to be taken about new and emerging technologies such as epigenetics, and there are still uncertainties about their risks and opportunities or it is not possible to predict the risks to which they will give rise. If the uncertainties and gaps in knowledge are so great as to preclude evidence-based decisions, the circle of people who participate in the participation process has to be expanded to include stakeholders (who represent the interests of business and civil society organisations) and/or "ordinary citizens". Public participation integrates societal norms, values, needs and perceptions into the process of negotiation, which can, firstly, help to introduce implicit knowledge and everyday experiences into the discourse, so broadening the range of perspectives and, secondly, lending the discourse and the decisions to which it leads a certain legitimation (cf. Hennen et al. 2004, p. 10).

3.3   The Limits of Public Participation

The ideas that have been discussed above could be read as a plea for public participation, at least under the auspices of TA activities on epigenetics. In this respect, a series of problems and unresolved questions arise that are also reflected in the scientific discourse about public participation—well beyond the field of participatory TA. Some are mentioned here by way of example.

One important question is that of the motivation(s) behind the organisation of public formats. Stirling (2008) identifies three different approaches that have already been outlined in the previous section: This is firstly, the 'normative' approach which applies public participation to research and innovation as a democratic principle. This thinking is founded on concepts such as RRI or citizen science which certainly in part have the aim of offering the people concerned a quasi-political right to be consulted ('empowerment'). Secondly, the 'instrumental' approach, which seeks to inform the public, and views dialogue between the scientific community and the public as a catalyst for controversies and conflicts that it ultimately aims to avert. And thirdly, the 'substantive' approach, which is intended to generate knowledge that will guide attempts to orient technology towards societal needs (as discussed in relation to participatory TA).

Precisely because the involvement of the public by scientists (working in TA) does not just serve the generation of knowledge, and certainly allows the possibility of using this involvement to intervene in and help shape public discourse, it is very important to declare these motivations in the interests of transparency. This could prevent misunderstandings and false expectations (e.g. with regard to the influence the results might have) among all concerned. In addition to this, it should be realised that the scientific community cannot be 'neutral'. For as soon as scientists involve the public, they inevitably intervene to a greater or lesser extent in public discourse in ways that are difficult to assess. Such interventions continue to be made, even though a public discourse cannot be controlled overall and follows its own rules, as is shown by the experience of discourses about other NESTs (Torgersen and Schmidt 2013).

Another urgent problem relates to the public's interest in taking part actively in the discourse about a NEST such as epigenetics. The other side of the 'Collingridge dilemma' is apparent here: the early stage of a technology's development not only opens up scope for it to be shaped, but has the disadvantage that there is still comparatively little knowledge available concerning its consequences. For this reason, the public is barely aware of the developments that are taking place and there are just a few concrete applications in prospect about which it would be possible to talk. Since, in practice, it is 'only' feasible to debate about the research agenda, active participation in discursive participation formats is often unattractive for citizens. This poses great challenges, in particular for direct participation by the public, and means the risk is run that a certain 'artificiality' will be perceived, the impact of which on the results and quality of the process is disputed (cf. Bogner 2012).

In the light of the present, media-led public discourse on epigenetics (a detailed survey of the current situation can be found in the article by Seitz and Schuol in the present volume), this argument may be moderated somewhat. Furthermore, the application of suitable quality criteria (e.g. Stirling 2008; Hennen 2012; Wehling 2012) could also help to soften its force—even if the question of the appropriateness of these criteria represents a field of research in its own right. In addition to this, there has been discussion of the lack of influence such procedures have on technology policy decisions, the possible instrumentalisation of participation to foster acceptance for mainstream science and technology policy, and the alleged falsification of the non-expert perspective during these processes due to the dominance of expert knowledge in their preparation (cf., among others, Wynne 2007; Hennen 2012; Irwin et al. 2013). It largely remains to be clarified how bridges can be built from the public discourse and its results back to the sphere of scientists, developers and political actors—i.e. how researchers and developers are sensitised to these debates and can be motivated to take account of the results in their work, as envisaged under RRI.

4   Conclusion

If epigenetics is looked at as an emerging (techno)science, it will be noted that its early stage of development is the ideal time to shaped it in line with the wishes and needs of society, and avert possible undesirable developments. This will require a comprehensive, open discourse between researchers, developers, political decision-makers, stakeholders and the interested public. In particular, the involvement of the interested public, which here means citizens without expertise or specific (economic) interests in epigenetics, appears especially important in this respect because they not only broaden the spectrum of perspectives with their implicit knowledge and everyday experiences but may, in certain circumstances, also lend the resulting decisions a certain legitimation (cf. Hennen et al. 2004, p. 10). This involvement can create an additional interface between society and political institutions which helps above all to open up political processes and improve or facilitate how state institutions respond to the concerns citizens articulate in controversies about technology (cf. Hennen et al. 2004, p. 8).

As far as this is concerned, it seems sensible to use and integrate the discourse on epigenetics that is taking place in any case (cf. Seitz and Schuol in the present volume) in appropriate forms, i.e. to draw lessons from its themes and the strands of the discussion. These lessons can both flow into the policymaking process within the TA framework and inspire academic discourse, e.g. on the ethics of epigenetics. However, the actors should be mindful of Paul Watzlawick's 1st axiom of communication theory ('One cannot *not* communicate', Watzlawick et al. 1967, p. 51). Any form of communication, even the mere publication of 'dry research results', will influence public discourse. This also means that the involvement of the public, regardless how it is organised, feeds back into current discourse. That is why it is advisable for the actors, above all in the scientific community, to be aware of their ability to shape that discourse and deploy it in a reflective fashion. However, engagement with the unresolved questions on public participation will become unavoidable, primarily in order to ascertain how the 'artificiality' of early discourses can be dealt with and how bridges can be built from these discourses back to the scientific community. Ultimately, it is irrelevant whether the public discourse is (actively) integrated into decision-making or not; nothing can prevent it from taking place. And since this discourse inevitably acquires political relevance, e.g. if it leads to conflicts and protests, it appears sensible not only for the actors to follow it attentively, but to use their own ability to shape it intelligently and reflectively. If they act in this way, it is highly likely that they will succeed in fostering an objective, constructive dialogue very much in the spirit of RRI, and that epigenetic research, with all its applications, will therefore contribute to the general welfare of society.

References

Bauer, M. W., Allum, N., & Miller, S. (2007). What can we learn from 25 years of PUS survey research? Liberating and expanding the agenda. *Public Understanding of Science, 16*, 79–95.

Bijker, W. E. (2013). Technology assessment—The state of play. In *Technology Assessment and Policy Areas of Great Transitions: Proceedings from the PACITA 2013 Conference in Prague.* Prague: PACITA.

Bird, A. (2007). Perceptions of epigenetics. *Nature, 447*(7143), 396–398.

Blech, J. (2010). *Gene sind kein Schicksal: Wie wir unsere Erbanlagen and unser Leben steuern können.* Frankfurt a.M.: S. Fischer-Verlag.

Bogner, A. (2012). The paradox of participation experiments. *Science, Technology and Human Values, 37*(5), 506–527.

Bütschi, D., & Nentwich, M. (2000). The role of participatory TA in the policy-making process. In L. Klüver, M. Nentwich, W. Peissl, H. Torgersen, F. Gloede, L. Hennen, J. van Eijndhoven, R. van Est, S. Joss, S. Bellucci, & D. Bütschi (Eds.), *European participatory technology assessment: Participatory methods in technology assessment and technology decision-making* (pp. 133–151). Brussels: European Commission.

Charlesworth, B., & Charlesworth, D. (2009). Perspectives: Darwin and genetics. *Genetics, 183*, 757–776.

Collingridge, D. (1982). *The social control of technology.* London: Open University Press.

Donges, P., & Imhof, K. (2001). Öffentlichkeit im Wandel. In O. Jarren & H. Bonfadelli (Eds.), *Einführung in die Publizistikwissenschaft* (pp. 101–133). Bern: Haupt.

Gerhards, J., & Neidhardt, F. (1991). Strukturen and Funktionen moderner Öffentlichkeit: Fragestellungen and Ansätze. In S. Müller-Dohm & K. Neumann-Braun (Eds.), *Öffentlichkeit, Kultur, Massenkommunikation* (pp. 29–89). BIS: Oldenburg.

Grunwald, A. (2010). *Technikfolgenabschätzung—eine Einführung,* 2nd edn. Berlin: Edition Sigma (Gesellschaft—Technik—Umwelt, Neue Folge 1).

Grunwald, A. (2012). Synthetische Biologie als Naturwissenschaft mit technischer Ausrichtung: Plädoyer für eine "Hermeneutische Technikfolgenabschätzung". *Technikfolgenabschätzung—Theorie und Praxis, 21*(2), 10–15.

Habermas, J. (1962). *Strukturwandel der Öffentlichkeit.* Neuwied/Berlin: Luchterhand.

Habermas, J. (1970). The scientization of politics and public opinion. In J. Habermas (Ed.), *Toward a rational society* (pp. 68–69). Boston: Beacon Press.

Handel, A., & Ramagopalan, S. (2010). Is Lamarckian evolution relevant to medicine? *BMC Medical Genetics, 11*(1), 73.

Hennen, L. (2012). Why do we still need participatory technology assessment? *Poiesis & Praxis, 9*(1–2), 27–41.

Hennen, L., Petermann, T., & Scherz, C. (Eds.). (2004). *Partizipative Verfahren der Technikfolgen-Abschätzung und parlamentarische Politikberatung: Neue Formen der Kommunikation zwischen Wissenschaft, Politik und Öffentlichkeit.* Berlin: Office of Technology Assessment at the German Bundestag (TAB), TAB working report no. 96.

Irwin, A., Jensen, T. E., & Jones, K. E. (2013). The good, the bad and the perfect: Criticizing engagement practice. *Social Studies of Science, 43*(1), 118–135.

Jasanoff, S. (2003). Technologies of humility: Citizen participation in governing science. *Minerva, 41*(3), 223–244.

Kegel, B. (2009). *Epigenetik: Wie Erfahrungen vererbt werden*. Cologne: DuMont Verlag.

Lefèvre, W. (2001). Jean Baptiste Lamarck. In I. Jahn & M. Schmitt (Eds.), *Darwin and Co: Eine Geschichte der Biologie in Portraits*, (Vol. 1, pp. 176–201). Munich: C.H. Beck.

Lehmkuhl, M. (2011). *Die Repräsentation der synthetischen Biologie in der deutschen Presse: Abschlussbericht einer Inhaltsanalyse von 23 deutschen Pressetiteln*. Berlin: Freie Universität Berlin, Institute for Media and Communication Studies.

Maletzke, G. (1972). *Psychologie der Massenkommunikation: Theorie and Systematik*. Hamburg: Hans Bredow Institute.

Negt, O., & Kluge, A. (1972). *Öffentlichkeit and Erfahrung: zur Organisationsanalyse von bürgerlicher and proletarischer Öffentlichkeit*. Frankfurt a.M.: Suhrkamp Verlag.

Rothstein, M. A., Cai, Y., & Marchant, G. E. (2009). The ghost in our genes: Legal and ethical implications of epigenetics. *Health Matrix Clevel, 19*(1), 1–62.

Stilgoe, J., Owen, R., & Macnaghten, P. (2013). Developing a framework for responsible innovation. *Research Policy, 42*(9), 1568–1580.

Stirling, A. (2008). "Opening up" and "closing down": Power, participation, and pluralism in the social appraisal of technology. *Science, Technology and Human Values, 33*(2), 262–294.

Swierstra, T., & Rip, A. (2007). Nano-ethics as NEST-ethics: Patterns of moral argumentation about new and emerging science and technology. *NanoEthics, 1*(1), 3–20.

Torgersen, H. (2013). TA als hermeneutische Unternehmung. *Technikfolgenabschätzung—Theorie and Praxis, 22*(2), 75–80.

Torgersen, H., & Schmidt, M. (2013). Frames and comparators: How might a debate on synthetic biology evolve? *Futures, 48*, 44–54.

von Schomberg, R. (2013). A vision of responsible research and innovation. In R. Owen, J. Bessant, & M. Heintz (Eds.), *Responsible innovation: Managing the responsible emergence of science and innovation in society* (pp. 51–74). Wiley: Hoboken.

Walk, H. (2013). Herausforderungen für eine integrative Perspektive in der sozialwissenschaftlichen Klimaforschung. In A. Knierim, S. Baasch, & M. Gottschick (Eds.), *Partizipation und Klimawandel—Ansprüche, Konzepte und Umsetzung* (pp. 21–35). Munich: Oekom Verlag.

Watzlawick, P., Beavin, J. H., & Jackson, D. D. (1967). *Pragmatics of human communication*. New York: W. W. Norton.

Wehling, P. (2012). From invited to uninvited participation (and back?): Rethinking civil society engagement in technology assessment and development. *Poiesis Praxis, 9*(1–2), 43–60.

Weinhold, B. (2006). Epigenetics: The science of change. *Environmental Health Perspectives, 114*(3), A160–A167.

Wynne, B. (2007). Public participation in science and technology: Performing and obscuring a political-conceptual category mistake. *East Asian Science, Technology and Society, 1*(1), 99–110.

Youngson, N. A., & Whitelaw, E. (2008). Transgenerational epigenetic effects. *Annual Review of Genomics and Human Genetics, 9*(1), 233–257.

Author Biography

Stefanie B. Seitz Dr. rer. nat. was senior scientist in the research area innovation processes and impacts of technology of the Institute of Technology Assessment and Systems Analysis (ITAS) at the Karlsruhe Institute of Technology (KIT) until 01/2016. She has published about the governance of manufactured nanomaterials, epigenetics and synthetic biology. Therefore, her research is focused on the question how society deals with the dilemma that new and emerging sciences and technologies poses and how the big promises can materialize and precaution with respect to uncertainties and potential risks can be maintained at the same. Moreover, she is interested in the potentials of public participation and the concepts of responsible research and innovation.

Contact: DBFZ – Deutsches Biomasseforschungszentrum gGmbH, Torgauer Str. 116, 04347 Leipzig, Germany.

State of the Public Discourse on Epigenetics

Stefanie B. Seitz and Sebastian Schuol

Abstract

Being part of molecular genetics, epigenetics deals with the regulatory mechanisms of gene-activity beyond the DNA sequence. The allegorical "being above the genes" of these mechanisms has revolutionised scientific theories and affected society's way of dealing with the stocks of genetic knowledge as well as the public discourse on it. To date there has hardly been any empirical research of these effects. In the following we present and evaluate three case examples. A media analysis explores the communicators' part. A discourse analysis explores the positions of recipients. There is an additional focus on the latter in form of an evaluation of a public event on epigenetics. Especially the aspect of self-responsibility plays role in the public discourse. So far the discourse proceeds very moderate and is primarily conducted by the media.

S.B. Seitz (✉)
DBFZ – Deutsches Biomasseforschungszentrum gGmbH,
Torgauer Str. 116, 04347 Leipzig, Germany
e-mail: stefanie.seitz@dbfz.de

S. Schuol
National Center for Tumor Diseases (NCT) Heidelberg,
Im Neuenheimer Feld 460, 69120 Heidelberg, Germany
e-mail: Sebastian.Schuol@med.uni-heidelberg.de

© Springer Fachmedien Wiesbaden GmbH 2017   109
R. Heil et al. (eds.), *Epigenetics*, Technikzukünfte, Wissenschaft und Gesellschaft / Futures of Technology, Science and Society,
DOI 10.1007/978-3-658-14460-9_9

1   Introduction

Epigenetics explores the molecular regulation mechanisms of gene-activity, working independently of the DNA. These mechanisms include signals which allegorically "are above the genes" and are potentially hereditary (see Walter and Hümpel in this volume). This special field of molecular genetics has fundamentally revolutionised our understanding of genes and the dogma according to which the phenotype of any living being is determined by its genes alone (see Schuol in this volume) has been invalidated. The insight that environmental factors and individual lifestyle considerably influence our health and possibly even our descendants' health over several generations is of high public interest and we have to face it in form of a public discourse (see Seitz in this volume).

However, so far there has hardly been any empirical research of the current state of this public discourse and public perception of epigenetics[1]—although academic reflection upon the societal aspects of epigenetic is gathering pace in the past years.[2] Some research shows that a discourse on the social consequences of epigenetics currently only takes place among a few expert circles, while the interested public only fulfil the role of audience (e.g. as those attending lectures or panel debates). Although epigenetics is a topic for the media (see media analysis), the voice of the citizens has not been documented there. In contrast to the discourse on genetic engineering, currently there are no indications of major public debates or protests. However, there are first interest groups pursuing the topic, such as Gen-ethisches Netzwerk e.V. (Gen-ethical Network) in Germany. Further evidence for a public discourse is the increasing media attention as well as the existence of blog contributions and discussion fora on this topic on the Internet. Thus, in the following three studies will analyse the state of the public discourse on epigenetics and its major topics: (1) an ad hoc media analysis, (2) an analysis of self-help literature on epigenetics including reactions by readers on the Internet, and (3) an evaluation of a public event on the topic.

---

[1]The only study we were able to identify was the pilot study of the ‚*Interdisciplinary Center of Epigenetics Science Studies*' at the OHSU under Prof. Deborah Heath, titled: "*The Emerging Epigenetic Mystique: Mapping Portrayals of Epigenetics in Social Media*" (https://college.lclark.edu/live/profiles/109-deborah-heath, last access March 2016).

[2]This is manifests itself in a growing number of opinion papers by bioethicists in life science magazines (e.g. Pickersgill et al. 2013; Waggoner and Uller 2015), activities like the this BMBF summer school or the Windsor Epigenetics Study Group (http://scholar.uwindsor.ca/emergingscholarspress/1, last access March 2016), but also in numerous popular science blogs.

2   Media Analysis

2.1   Ad Hoc Media Analysis as an Indicator of the State of the Public Discourse

The analysis of media contents is a crucial part of social-scientific discourse analyses as they have developed following Michel Foucault's statements on the concept of discourse (see Klemm and Glasze 2005). In the sense of a "prognostic approach" (Früh 2004, p. 41f.), the analysis of media contents may provide a basis to study its effect on recipients, for example for increasing their knowledge of a certain field (Maurer and Reinemann 2006). Although the here presented media analysis is only a simple one, due to limited resources, and does not meet the standards of a classical media analysis of communication studies, it is suitable for providing first insights into the topics of media coverage on epigenetics and to estimate on its frequency.[3]

2.2   Medial Presence of the Topic

The first result of this analysis is that the reporting on epigenetics starts already during the early stages of this scientific sub-discipline.[4] During this early stage, up to 2004, the contributions present epigenetics as a new field of research, explain its novelty as well as the "revolutionary new thinking" resulting from epigenetic insights, and in interviews the researchers tell about their visions.[5] After a decline

---

[3]For the here presented results, contributions under the keyword "epigen ∼ " were analysed most of all in the German-speaking, but also international, daily and weekly press [e.g. FAZ, SZ, FR, Welt, TAZ, Tagesspiegel, Focus, Spiegel, NZZ (CH), Standard (AT), Guardian (GB)], and a variety of Internet sources was assessed (e.g. Scinexx, search via Google). In the period 1999–2013 108 contributions could be identified, which is an attempt at giving the complete output.

[4]The earliest accessible article is "VERERBUNG: Gene sind nur Marionetten" in Der Spiegel of 04.11.2002. We were able to investigate some more which were not freely accessible; according to the headlines, they must be filed under "scientific progress and the opportunities of epigenetics" (see Sect. 2.3). The earliest article researched GENIOS is of 29.06.1994 and was published by FAZ, although it deals first of all with gene therapy.

[5]Examples: "Übergenom—Der Lamarck-Code: Ein neues Humangenomprojekt beginnt" (FAZ, 08.10.2003) and "Krankheiten vor Ort verstehen: Jörn Walter, 52, Professor für Epigenetik an der Universität des Saarlandes in Saarbrücken, über die Veränderlichkeit des Erbguts" (Der Spiegel, 08.11.2010).

of the coverage between 2004 and 2007 the number of contributions rises again from 2008 on, and also the topics become more varied. At this time there also appear the first critical voices, even if they are still exceptions.[6] In the most cases the focussed coverage is triggered by spectacular scientific publications (e.g. Rice et al. 2012 on homosexuality), but also by the publication of popular-scientific books on the topic whose contents are reviewed and reflected on.

2.3   Guiding Themes of the Coverage of Epigenetics

The contents of the contributions under analysis can be classified by four guiding themes. By far the most contributions (ca. 50 %) deal with the topic of "scientific progress and the opportunities provided by epigenetics". They inform about new insights gained by research, e.g. the discovery of new epigenetic mechanisms, often pointing out to the thus connected hopes concerning the therapy and diagnostics of widespread diseases (such as cancer, diabetes or Alzheimer).[7] These contributions have a mostly optimistic and technology-supporting tendency while potential risks are hardly discussed. They belong to the tradition of classical, neutral science journalism, in the context of which the authors do not take a position, as we also find it when it comes to the coverage of other innovative technologies such as nanotechnology (see Haslinger et al. 2012, p. 33), and they are often found on the science pages of daily newspapers.

In recent years the contributions discuss the "overcoming of the dogma of genetics" as a guiding theme (ca. 30 %) and a "victory over the genes" has been declared (cover story of Der Spiegel August 9th, 2010). The starting point of the reporting on epigenetics is the insight of epigenetic research according to which—to put it simply—an individual's fixedly determined fate is not that "inevitable", after all, but—as it is suggested—can be prevented by way of positively changing one's own environment.[8] Accordingly, there we find a number of "fitness tips for your genes", even if in most cases they are only commonplaces such as avoiding

---

[6]Example: "Mutti ist an allem schuld—Der neue Irrglaube an die Machbarkeit des Menschen." (Tagesspiegel, 24.04.2013).

[7]Examples: "Ribonukleinsäure: Kleine Schnipsel, große Wirkung" (Der Spiegel, 01.01.2003) and "Stumme Zeugen—Stillgelegte Gene sind am Entstehen von Krebs beteiligt" (Tagesspiegel, 16.06.2010).

[8]This is clearly suggested by some articles (e.g. "Klüger, gesünder, glücklicher: Wie wir unser Erbgut überlisten können", Der Spiegel, 09.08.2010), however already the articles themselves express reservations and relativizations.

stress, maintain friendships, sports, a healthy diet and the likes. These contributions refer to the nature vs. nurture debate and make self-responsibility a topic of discussion: Now the individual may no longer refer to its "genetic fate" but has its own "well-being" in his/her own hands (again). This kind of contributions is mostly found in the big weekly magazines, probably due to their standard service of offering advice.

The contributions under the guiding theme of the "memory of the genes" are the flipside of the previous category, but they are not as frequent (ca. 15 %). They discuss harmful environmental influences on "epigenetics" that are out of the individual's control, e.g. negative influences on the unborn child, traumatic experiences or environmental toxins.[9] There we find, although only in single cases, criticism of the expectations connected to medical applications, but also of the suggested possibility of oneself purposefully influencing one's own gene regulation. There, also the topic of responsibility is discussed, in particular responsibility towards one's own offspring, but also concerning the state's obligations to protect.

A small group of contributions (ca. 5 %) deals particularly with those insights of epigenetic research as being connected to sexuality and partnership.[10] Topically they belong to the first guiding theme, however they stand out due to being particularly detailed and more interpretative. These themes draw the attention of readers, which potentially leads to growing circulation. Probably that is why they are often referred to. Related to the scientific relevance of the insights, disproportionally often there is a focussed reporting. Due to the aspect of sensation, it is probable that the topics of this kind will be taken up by the public discourse (on the case of homosexuality see Matern in this volume).

3 Discourse Analysis

3.1 Epigenetics in Self-help Literature

The establishment of epigenetics in the German-speaking countries was accompanied by a number of popular-scientific books (Bauer 2002, 2008; Blech 2010; Huber 2010; Kegel 2009; Spork 2009) which, beyond communicating the complex

---

[9]Examples: "Schwangere im Stresstest—Frauen stellen die Weichen für die Zukunft" (Focus, 29.11.2010), "Gewalt verändert Genom der Kinder" (FAZ, 20.07.2011) and "Rauchen bringt Genchemie durcheinander" (Scinexx, 04.04.2011).

[10]Examples: "Forscher erklären Mysterium der Homosexualität" (Welt, 11.12.2012) and "Die Chemie der Monogamie" (SZ, 04.06.2013).

expert knowledge, have an advice-giving function, thus decisively influencing the public discourse as far as to concrete action. Although these books differ in details, they have common features which will be discussed in the following.

Firstly, all authors identify an epochal turn in biological thinking. After an epoch of genetic determinism the tide is turning and genetic plasticity is now a major topic. Thereby all authors refer to the Human Genome Project as a turning point and locate epigenetics in the following era of post-genomics.

Secondly, it is argued that molecular processes can be actively controlled. All authors emphasize the practical application of epigenetics, while stating that the individual lifestyle has the status of a bio-technology (Schuol 2015). The supreme objective health is discussed particularly in the context of life style disease like e.g. metabolic syndrome (Schuol 2014).

Thirdly, the authors deduce from what has been said above a new responsibility. Notably this new responsibility is not, as one might assume, communicated as a burden but as an achievement in the positive sense. The reason for this is an emancipator one; the new knowledge of gene-regulation refutes genetic determinism and enables autonomous actions. It is also notable that the authors discuss a variety of social, political and individual responsibilities.

3.2   Epigenetics from the Recipients' Point of View

Thus far, the topics of self-help literature are mostly congruent with the findings of the media analysis. However, looking at the debates among readers an interesting change is striking. In blogs and smaller publications the topics of self-help literature are taken up, discussed and, while doing so, transformed. That the act of reception is an activity of its own becomes obvious by the topic of responsibility. Whereas the readers repeat the other topics without substantial change, the topic of responsibility changes into self-responsibility. This happens both among critics (see Peuker 2010) and supporters (see Strunz 2012) of the self-help literature on epigenetics. This transformation demonstrates that the discourse on epigenetics is no isolated entity but that its topics are interpreted in the light of the already existing discourses on self-care and self-optimization. Since self-responsibility ("Eigenverantwortung" in German) is a "young" term, its interpretative framework can be located and the meaning of the term can be explained more precise. In Germany the term self-responsibility has been used as a political tool in the context of the reform of the social security system (Agenda 2010). The goal of these reforms was to increase the citizens' autonomy, i.e. their capability of self-regulation, for the purpose of easing the system of social welfare. Thereby responsibility meant first of all self-responsibility. Since the readers simplified the term in this way, this allows conclusions on their socio-economic background. It is well known that health education in the context of the reform of the health system was especially in the educated middle class successfully (see Niewöhner 2010). This class took up the new insights and integrated them into its own thinking, in order of acting preventive in an autonomous way (see Mathar 2010, p. 185ff.). The high level of education among the recipients is also confirmed by the following analysis of public events. Vice versa it can be assumed that the less educated classes are underrepresented in the discourse. Due to this selectivity, generalizing statements on the public discourse on epigenetics and its topics must be put into question.

There is another reason for criticising the focus on self-responsibility. Due to the interpretation of lifestyle as a bio-technology the restricted view of the individuals' epigenetic responsibility seems plausible (Schuol 2015). However, at a closer look this interpretation becomes dubious since it ignores social influences, supports already existing social injustices and tends to overburden the individual because of leaving operational limits undetermined (Schuol 2014). In this context of prevention self-responsibility can only have the status of a sub-responsibility. This holds even more if, as it is assumed in the light of environmental epigenetics, the inequalities become epigenetically hardwired and the social status will be passed further to the following generations biologically (Niewöhner 2011, 2013).

3.3 Spiritual Topics in the Discourse on Epigenetics

The public discourse on epigenetics is characterised not only by scientific actors and their visions. Currently epigenetics plays a crucial role in the context of the so called New Biology, a spiritual movement which is closely connected to the New Age movement (Ellison 2010). Bruce Lipton, one of the spiritual leaders of this movement worked out the fundament in his book "Biology of Belief" published in 2006. Although it moves away from the traditional understanding of health this book may also be filed under the category of self-help literature since it is considered very important in the relevant discussion fora.

Lipton's starting point is the historical struggle between materialism and idealism and the question if matter determines the spirit or vice versa. Here, the above sketched thesis of controllability becomes radically extended: It is assumed that the state of mind determines biological existence up to the molecular level of gene regulation and both spiritual and physical health can be mentally controlled. Here, the main thesis of Historical Materialism, according to which conditions determine consciousness, is reversed into its opposite.

In the esoteric context, epigenetics gains a spiritual dimension that appears somewhat problematic. Indeed, the health-supporting effect of belief and hope is discussed even in the academic context (see Frewer et al. 2010). However, exaggerated idealism may soon change to the opposite and becomes a burden namely if self-responsibility changes into personal guilt: Whoever regards exclusively himself as the author of his fortunes blames vice versa consequently himself for (his) misfortunes. The thesis is especially with regard to the health problematic since the means to ensure its own health never are completely in the hands of individuals—partly for reasons of complexity, partly because of stochastically occurring random processes. In the strict sense even in the light of epigenetics spiritual and physical health remains unavailable for individuals since they can only control wider health conditions.

4   Evaluation of a Public Event

4.1   Background of the Event

The public event which shall be evaluated in the following was the concluding event of the BMBF—(Federal Ministry of Education and Research) funded week-long interdisciplinary summer school on "Epigentics—Ethical, Legal and Social Aspects" (September 15th–20th, 2013). It happened in the evening of September 20th, 2013, in the lobby of ITAS in Karlsruhe and was called: "Epigenetics: Can We Influence Our Genes?" It was officially opened by an introductory lecture on epigenetics. Then the guests were given the opportunity to discuss the matter with the scientists who presented the topics of the summer school (see the contributions by Bode, Fündling, Jahnel, Schuol and Seitz in this volume) at posters which presented the topics of the conference in a simple language. Then there followed a panel debate with scientists from the legal and natural sciences as well as from the humanities.

As up to now it has not been possible to research any documentation of a public event on epigenetics, this one was purposefully scientifically accompanied. Due to the small number of participants, its evaluation cannot provide any statistically relevant results, however at the qualitative level it provides valuable insights into the state of the public discourse. As a basis for the evaluation there served questionnaires asking about demographic data (age, gender, level of education),

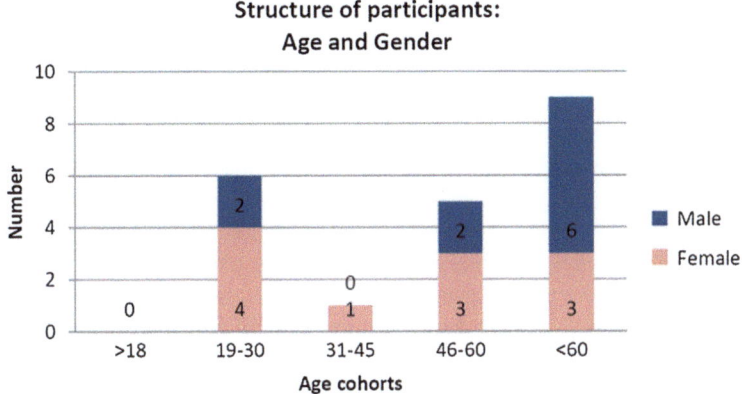

Fig. 1 Structure of participants by age cohorts and gender; n = 21. Own figure

interest, state of knowledge and informational behaviour as well as about the event itself.[11] The questionnaires were voluntary and anonymous. Furthermore, the event was subject to a participating observation, and the scientists present during the event were interviewed about their discussions with the guests.

The structure of the participants is presented in Fig. 1 and is, according to our experiences, in line with the typical structure of public events on scientific/technological issues.

4.2 The Audience's State of Knowledge of the Topic of Epigenetics

Despite common prejudices among scientists, the extended quantitative and qualitative surveys of recent years have repeatedly shown that the population of western industrialised countries—in particular Germany—have a rather positive attitude towards technology and are capable of differentiate their judgments between the different cases (see Eurobarometer 2010, 2013). Probably those attending the here analysed event rather belonged to those showing a rather positive attitude towards technology and the sciences, due to the venue it must be ruled

---

[11]The complete number of questionnaires was returned and is in the authors' hands.

out that their presence was coincidental.[12] For example, in the questionnaire 71 % of the participants stated that they had been familiar with the term "epigenetics" already before, and 48 % stated to know even "much" about it. The answers to comparable questions in the course of representative surveys e.g. on how well synthetic biology is known, only 18 % of the Germans asked were familiar with the term (Eurobarometer 2010). Thus, it must be assumed that those attending the event are not necessarily representative for the German population. The questions asked by the guests showed that not all of them were familiar with the term "epigenetics" as well as the concept behind it. Among others, they asked questions rather concerning the field of genetics, but also concerning nanoparticles (which might be due to the scientific background of one of the experts). Furthermore, almost all guests stated that they considered the issue of epigenetics to be important and that they would like to know more about it. They were also asked about their preferred sources of information: Independently of age and gender, print media (81 %) and the Internet (76 %) were preferred. This finding illustrates the significance of Internet sources, which is still insufficiently taken into consideration by current media analysis.

4.3   Major Issues During the Discussion with the Audience

The focal points of the evening were determined by the framework of the event, i.e. the topics presented by the posters were predominant also later for the panel debate. However, the participating observation as well as the evaluation of the questionnaires showed that the guests were particularly moved by certain issues and that they definitely introduced some of their own.

Predominantly there were questions for a better understanding of epigenetics. What exactly is epigenetics? What is the difference to genetics? What is new about it? Some guests showed a sceptical attitude towards the novelty of epigenetics as a new scientific field and were of the opinion that it was "artificially boosted" or "unnecessary". On the whole, the guests were most of all interested in the "interdisciplinary research of epigenetic mechanisms and interdependencies when it comes to the causes of illnesses and the development of therapies", which is also

---

[12]The event was announced in the following ways: (1) An invitation flyer inserted in the Badische Neuste Nachrichten one week before the event, (2) via several e-mail lists (Volkshochschule Karlsruhe, KIT Public Relations and internal mailing lists) and (3) Facebook.

mirrored by the questionnaire. They were very interested in how this knowledge could be transferred to everyday life and in getting to know which consequences it might have for their lives or their behaviour. Also during the discussions with the scientists they asked e.g. about hints for "correct" behaviour during pregnancy or about explanations for illnesses in the closer family environment. Approaching a new topic by way of being personally concerned or affected is a phenomenon which is well-known in social research (see Fleischer et al. 2012).

Another important aspect for the guest was the issue of responsibility: On the one hand one's own responsibility resulting from the potential trans-generationality of epigenetic mechanisms, and on the other hand the responsibility of the government. In this context, epigenetics is embedded into the overarching topics of food safety and consumer protection. For example, the currently running debate on the omnipresence of plastics and the plasticizers they contain were repeatedly addressed. In this context several guests expressed their criticism of the regulations for evidently potentially health-damaging chemicals in food, such as Bisphenol A (see Jahnel in this volume). They considered the existing regulations to be too soft. Concerning this, they asked "What are the insights of epigenetics good for, after all, if dangers (such as BPA) which have already been identified upon existing scientific knowledge or even upon common sense are not regulated?" A great number of those attending expressed their serious dissatisfaction with current consumer policy and believed to be "fooled".

4.4 Participation in the Public Discourse

In the questionnaire all guests stated that they believed the issue of epigenetics to be socially relevant, and every second one believed to be personally concerned, younger guests (19–45 years old, 71 %) feeling more concerned than older guests (over 46, 46 %). This suggests that the social discourse should be intensified; indeed in the questionnaire all guests agreed with this. Opinions were much different concerning the question of what the discourse was supposed to be like. For example, 15 % were of the opinion that such a discourse should only happen among experts, 45 % on the other hand were of the opinion that the discourse should happen in the media. The majority (70 %) would like all groups of society to participate in the discourse, with only 30 % being of the opinion that the citizens should participate directly (multiple answers were possible).

Given the growing popularity of deliberation and the various ways of citizens' participation (see Hahn et al. 2014) this may come as a surprise, but it is also confirmed by the recent Eurobarometer survey (2013). From our own field research in this field we know that, when it comes to play an active role with processes of dialogue, citizens are uncertain about their contributions, as they are "lay people". This question, about the value of citizens' participation in social debates, was indeed raised by the guests themselves, but answering it led deeply into the heart of the scientific debate on the participation of lay people and is discussed elsewhere (see Seitz in this volume).

5   Conclusions

These three analyses provide a first look at the current situation of the public discourse on epigenetics in the German-speaking countries. As a conclusion it must be stated that, although just in its beginnings, this discourse is already happening. It must be noted in this context that those parts of the population as participating in it are not representative; usually they are well educated and/or personally concerned.

The analysis of the contents of relevant non-fictional books and contributions in the print media (incl. the latter's online appearances) make the following guiding themes obvious: (1) reports about the scientific progress and the opportunities of epigenetics, (2) overcoming genetic determinism, (3) the memory of genes, as well as (4) epigenetics and sexuality (decreasing according to frequency). This shows which topics and narratives are fed into the discourse.

However, the effect of the media is controversially discussed among the scientific community, and on the whole it is only insufficiently understood. As shown by Bonfadelli and Friemel (2011), it cannot be predicted what is to which degree perceived by media users and which consequences it might have for their behaviour. It is assumed that media contents determine the framework of understanding (framing) and influence attention (priming). It is thus difficult to determine to which degree this happens. Thus although we must assume a sustainable effect of topics raised by the media, at the same time the analysis of the recipient side shows that there is no comprehensive determination by the media. This finding is also congruent with the current state of media theory, which considers the idea of media consumers being one-sidedly determined by media producers to be invalidated (Maletzke 1998, p. 55). On the contrary, the recipients of media transform the topics (e.g. reducing responsibility issues to self-responsibility), connect them to the current controversies and add their own issues.

On the whole, it seems as if the public discourse on epigenetics is by far not as controversial and anxiety-driven as that on genetics/genetic engineering, for example. Currently the public rather shows a general interest in epigenetics—to quench their thirst for knowledge and while expecting good advice. Possibly epigenetics are still predominantly perceived as a field of research—without any actual implementations which might trigger fear and conflicts. Realised and predicted implementations of epigenetics are found mostly in the medical realm which traditionally meets more approval even if the risk profile is less favourable (Fleischer et al. 2012). However, this description covers only the situation at the time of our observations and does not allow for predictions. As has been shown by technology discourses of the past, one single incident may suddenly completely change the direction of discourses and change approving interest into protest—with the appropriate effects on the acceptance of implementation.

Given the explosive nature of the topics and the selective participation, finally it must be considered if the public discourse shall be additionally stimulated and provided with input by the sciences. Participative technology assessment (see Seitz in this volume) may contribute to a normalisation of the discourse, by the different social spheres encountering each other and exchanging their opinions. This way scientific-technological developments may be influenced by society's needs on the on the one hand, but just the same the latter may also influence governance measures, quite in the sense of responsible research and innovation and as a contribution to sustainable decision-making concerning the frames of research and implementation (see Stilgoe et al. 2013).

References

Bauer, J. (2002). *Das Gedächtnis des Körpers: wie Beziehungen und Lebensstile unsere Gene steuern*. Eichborn: Frankfurt a.M.

Bauer, J. (2008). *Prinzip Menschlichkeit—Warum wir von Natur aus kooperieren*. München: Heyne.

Blech, J. (2010). *Gene sind kein Schicksal: Wie wir unsere Erbanlagen und unser Leben steuern können*. Frankfurt a. M.: S. Fischer Verlag.

Bonfadelli, H., & Friemel, T. N. (2011). *Medienwirkungsforschung*. Konstanz: UVK.

Ellison, K. (2010). New age or "new biology"? *Frontiers in Ecology and the Environment, 8*, 112.

Eurobarometer. (2010). Special eurobarometer 341/Wave 73.1—biotechnology. Report. Fieldwork: January–February 2010. Brüssel: Europäische Kommission. http://ec.europa. eu/public_opinion/archives/ebs/ebs_341_en.pdf. Last accessed March 2016.

Eurobarometer. (2013). Special eurobarometer 401—Responsible research and innovation (RRI), science and technology. Report. Fieldwork: April–May 2013. Brüssel: Europäische Kommission. http://ec.europa.eu/public_opinion/archives/ebs/ebs_401_en.pdf. Last accessed March 2016.

Fleischer, T., Haslinger, J., Jahnel, J., & Seitz, S. B. (2012). Focus group discussions inform concern assessment and support scientific policy advice for the risk governance of nanomaterials. *International Journal of Emerging Technologies and Society, 10*(1), 79–95.

Frewer, A., Bruns, F., & Rascher, W. (Eds.). (2010). *Hoffnung und Verantwortung Herausforderungen für die Medizin.* Würzburg: Jahrbuch Ethik in der Klinik (JEK). Bd. 3.

Früh, W. (2004). *Inhaltsanalyse. Theorie und Praxis.* Konstanz: UVK.

Hahn, J., Seitz, S. B., & Weinberger, N. (2014). What can TA learn from 'the people'? A case study of the German citizens' dialogues on future technologies. In T. C. Michalek, L. Hebakova, L. Hennen, C. Scherz, L. Nierling, & J. Hahn (Eds.), *Technology assessment and policy areas of great transitions* (pp. 165–170). Prag: Technology Centre ASCR.

Haslinger, J., Hauser, C., Hocke, P., & Fiedeler, U. (2012). *Ein Teilerfolg der Nanowissenschaften? Eine Inhaltsanalyse zur Nanoberichterstattung in repräsentativen Medien Österreichs, Deutschlands und der Schweiz.* Wien: ITA-manuscript (12-04).

Huber, J. (2010). *Liebe lässt sich vererben. Wie wir durch unseren Lebenswandel die Gene beeinflussen können.* München: ZS Verlag.

Kegel, B. (2009). *Epigenetik: Wie Erfahrungen vererbt werden.* Köln: Dumont.

Klemm, J., & Glaze, G. (2005). Methodische Probleme Foucault-inspirierter Diskursanalysen in den Sozialwissenschaften. Tagungsbericht: "Praxis-Workshop Diskursanalyse", 6(2). http://pub.uni-bielefeld.de/publication/2404700. Last accessed March 2016.

Lipton, B. H. (2006). *Intelligente Zellen: Wie Erfahrungen unsere Gene steuern.* Burgrain: Koha Verlag.

Maletzke, G. (1998). *Kommunikationswissenschaft im Überblick. Grundlagen, Probleme, Perspektiven.* Opladen: Westdeutscher Verlag.

Mathar, T. (2010). *Der digitale Patient. Zu den Konsequenzen eines technowissenschaftlichen Gesundheitssystems.* Bielefeld: Transcript (VerKörperungen, 10).

Maurer, M., & Reinemann, C. (2006). Learning versus knowing: Effects of misinformation in televised debates. *Communication Research, 33*(6), 489–506.

Niewöhner, J. (2010). Über die Spannung zwischen individueller und kollektiver Intervention. Herzkreislaufprävention zwischen Gouvernementalität und Hygienisierung. In M. Lengwiler & J. Madarázs (Eds.), *Das präventive Selbst. Eine Kulturgeschichte moderner Gesundheitspolitik* (pp. 307–324). Bielefeld: Transcript (VerKörperungen, 9).

Niewöhner, J. (2011). Epigenetics: Embedded bodies and the molecularisation of biography and milieu. *BioSocieties, 6*, 279–298.

Niewöhner, J. (2013). Neue Sozialhygiene oder lokale Biologie? *Gen-ethischer Informationsdienst, 220*, 13–15.

Peuker, B. (2010). Übernimm Verantwortung für dein Genom. *Gen-ethischer Informationsdienst, 199*, 37–39.

Pickersgill, M., Niewöhner, J., Müller, R., Martin, P., & Cunningham-Burley, S. (2013). Mapping the new molecular landscape: Social dimensions of epigenetics. *New Genetics and Society, 32*(4), 429–447.

Rice, W. R., Friberg, U., & Gavrilets, S. (2012). Homosexuality as a consequence of epigenetically canalized sexual development. *The Quarterly Review of Biology, 87*(4), 343–368.

Schuol, S. (2014). Kritik der Eigenverantwortung: Die Epigenetik im öffentlichen Präventionsdiskurs zum Metabolischen Syndrom. In V. Lux & T. Richter (Eds.), *Vererbt, codiert, übertragen: Kulturen der Epigenetik* (pp. 271–282). Berlin: De Gruyter.

Schuol, S. (2015). Lebensstil als Biotechnik? Zur Erweiterung des Genbegriffs durch die Epigenetik. In R. Ranisch, M. Rockoff, & S. Schuol (Eds.), *Die Selbstgestaltung des Menschen durch Biotechniken (in Arbeit)*. Francke: Tübingen.

Spork, P. (2009). *Der zweite Code. Epigenetik—oder wie wir unser Erbgut steuern können.* Reinbek: Rowohlt.

Stilgoe, J., Owen, R., & Macnaghten, P. (2013). Developing a framework for responsible innovation. *Research Policy, 42*(9), 1568–1580.

Strunz, U. (2012). Epigenetik ist Eigenverantwortung. NEWS,13.09.2012 at "forever-young-Strunz"-Homepage. http://www.strunz.com/news.php?newsid=1910. Last accessed March 2016.

Waggoner, M. R., & Uller, T. (2015). Epigenetic determinism in science and society. *New Genetics and Society, 34*(2), 177–195.

Author Biographies

Stefanie B. Seitz Dr. rer. nat. was senior scientist in the research area innovation processes and impacts of technology of the Institute of Technology Assessment and Systems Analysis (ITAS) at the Karlsruhe Institute of Technology (KIT) until 01/2016. She has published about the governance of manufactured nanomaterials, epigenetics and synthetic biology. Therefore, her research is focused on the question how society deals with the dilemma that new and emerging sciences and technologies poses and how the big promises can materialize and precaution with respect to uncertainties and potential risks can be maintained at the same. Moreover, she is interested in the potentials of public participation and the concepts of responsible research and innovation.

Contact: DBFZ – Deutsches Biomasseforschungszentrum gGmbH, Torgauer Str. 116, 04347 Leipzig, Germany.

Sebastian Schuol is scientific coordinator of the EURAT-project (Ethical and Legal Aspects of Whole Genome Sequencing) which is located at the University of Heidelberg. After studying philosophy and molecular genetics he was a fellow at the DFG Research Training Group "Bioethik" at the International Center for Ethics in the Sciences and Humanities (IZEW) in Tübingen and completed his dissertation in philosophy of biology on the theoretical and practical impact of epigenetics on the concept of the gene. He has a special interest in philosophy of science, theory of biology and bioethics. So far he has published two articles on epigenetics (Kritik der Eigenverantwortung: Die Epigenetik im öffentlichen Präventionsdiskurs zum Metabolischen Syndrom, Berlin 2014; Der Lebensstil als Biotechnik? Zur Erweiterung des Genbegriffs durch die Epigenetik, Tübingen 2015).

Contact: National Center for Tumor Diseases (NCT) Heidelberg, Im Neuenheimer Feld 460, 69120 Heidelberg, Germany.

Epigenetics—New Aspects of Chemicals Policy

Jutta Jahnel

Abstract

Epigenetics is the study of the natural processes that regulate the differentiation of cells and tissues, and are of significance in the development of organisms. This means it is able to offer a specific perspective on how various factors and stressors in the environment control gene activity. For instance, chemicals such as hormonally active 'endocrine disruptors' leave behind traces in the epigenetic code that not only trigger illnesses, but can also be passed on from generation to generation. These epigenetic mechanisms of action are not taken sufficiently into consideration in the use and regulation of chemicals. Although the evaluation of hormone disrupting effects due to endocrine disruptors is covered by various product-specific European regulations, neither an unambiguous regulatory definition nor specific toxicological testing strategies have been in place to date. As a result of this, the protective measures provided for in the existing legislation remain vague. It has not been possible up until now for the state to monitor the use of these substances in everyday products. How society deals with the consequences of the influence epigenetic mechanisms have due to anthropogenically conditioned changes in the environment constitutes an interdisciplinary challenge for the academic community, policymakers and ethicists. The European Parliament has adopted a resolution on the protection of public health from endocrine disruptors in which it demands precautionary action because risks to the environment and human

J. Jahnel (✉)
KIT—The Research University in the Helmholtz Association,
Institute for Technology Assessment and Systems Analysis (ITAS),
Karlstraße 11, 76133 Karlsruhe, Germany
e-mail: jutta.jahnel@kit.edu

© Springer Fachmedien Wiesbaden GmbH 2017
R. Heil et al. (eds.), *Epigenetics*, Technikzukünfte, Wissenschaft und
Gesellschaft / Futures of Technology, Science and Society,
DOI 10.1007/978-3-658-14460-9_10

health cannot be ruled out. This means epigenetics has arrived as a topic on the current political agenda in the context of the impact assessment and regulation of substances with epigenetic mechanisms of action.

1   Introduction: The Significance of Epigenetics

Epigenetics is a field of research that studies the mechanisms by which the observable characteristics of a living organism, what are known as its 'phenotype', are formed. Apart from their genetic constitution ('genotype'), the particular role of environmental factors in the development and formation of individual organisms is taken into account. The meaning of the term 'epigenetics' has changed over time. Inspired by the concept of epigenesis as a dynamic developmental system that forms 'individual organisms from initially undifferentiated cellular tissues', Conrad Waddington describes 'epigenetics' as the interactions of genes with their environment, the entirety of which produce the 'phenotype'. Today, the term 'epigenetics' is understood above all within a molecular biological paradigm (cf. Schuol in the present volume). Here, 'epigenetics' denotes the inheritable molecular biological information that is stored by cells and passed on to their daughter cells, but is not contained in the sequence of the deoxyribonucleic acid (DNA) itself. As a result of this, there is a formal separation between genetic and epigenetic information. What could be called a broad reading frame that controls genetic activity and acts like an 'on' and 'off' switch for gene expression is laid on top of the genetic code determined by the DNA. As the manifestation of genetic information in the phenotype, gene expression occurs at different levels. During transcription, DNA is read and reproduced by a ribonucleic acid (RNA) molecule. A specific section of DNA therefore serves as the template for the synthesis of a new RNA strand. During translation, a protein is then synthesised from the RNA. This gives rise to various possibilities for the epigenetic control of gene activities (cf. Rothstein et al. 2009).

Genes are (de)activated epigenetically by biochemical processes, for the most part the methylation of DNA or changes in its three-dimensional structure (methylation/acetylation of histone proteins). Both mechanisms prevent the gene from being read and accordingly have effects at the level of transcription. However, gene activity can also be controlled by the disruption of translation. Epigenetic markings therefore regulate transcription factors, but also post-transcriptional modifications to RNA, and are fundamentally reversible. This means the significance of epigenetics is quite plain: It serves to control natural processes that regulate cell and tissue differentiation, and is responsible for the development of individual organisms.

2 The Influence of Chemicals on Epigenetic Markings and Possible Consequences

While the blueprints for gene products are stored largely stably in the genetic information and, apart from mutagenic influences, are not subject to direct modifications due to environmental effects, the epigenetic code can be altered by chemicals. This has been demonstrated both in microorganisms, plants and animals, and in human beings (Weinhold 2012). If the epigenetic marking in body cells changes, this influences the physiology of the individual who has been exposed and therefore intra-generational development. The implication is that prenatal and/or postnatal exposure to chemicals can cause phenotypic changes in human beings over the further course of their lives (Jirtle and Skinner 2007). If there are changes to the epigenetic marking in the germ cells, this is referred to as epigenetic transgenerational inheritance (Skinner et al. 2010; Skinner 2011; Schug et al. 2011).

Epigenetic changes can cause specific diseases. These mechanisms of action play a significant role in, among other things, the development of cancer, autoimmune conditions, asthma, diabetes and heart conditions (Feinberg 2007; Owen and Segars 2009; Newbold et al. 2009; Tang et al. 2012; Porta and Lee 2012). However, complex pathological phenomena such as infertility, obesity and neurological defects can also be connected with modified epigenetic marking (Nilsson et al. 2012; Leranth et al. 2008). Nevertheless, epigenetic changes are not in principle associated with adverse effects. Even where it is proven that methylation of the DNA has taken place, this therefore does not inevitably entail health effects. At the same time, it is still not at all possible to assess the potential health damaging impacts that can be caused unintentionally by different fields of technology, and they have also been disregarded to a very great extent in the governance of these technologies so far.

With regard to the evaluation of the production and usage of chemicals in consumer products and in the environment, the focus is moving to epigenetic modifications. Generally, chemicals may only be used once they have been tested and registered in a procedure stipulated by legislation. In Europe, chemicals have to be registered in accordance with the provisions of the Regulation on the Registration, Evaluation, Authorisation and Restriction of Chemicals (Regulation

(EC) 1907/2006, REACH) if the amount produced is one tonne or more per year. Following registration, they are included in the Community Rolling Action Plan (CoRAP) that specifies the substances to be evaluated. Such an evaluation can, among other things, lead to a substance restriction or authorisation process. Depending on the quantity that is produced, it is necessary to draw up either a registration dossier or an extensive Chemical Safety Report. Apart from the substance's physical and chemical properties, it is prescribed that the dossier or report should contain details of the ecotoxicological and toxicological tests that are carried out. The 'risk assessment' of chemicals is a four-stage process that consists of exposure assessment, hazard identification, dose-response assessment and risk characterisation (NRC 1983). However, possible epigenetic mechanisms of action attributable to chemicals have not yet been taken into consideration explicitly under this approach to date. Although there are special provisions concerning the risk characterisation of carcinogenic and genotoxic chemicals that trigger changes in DNA, no adequate strategies for the assessment of carcinogenic substances with epigenetic mechanisms of action are available. Over the last few years, though, it has been possible to observe marked scientific and political interest in the risk assessment of what are known as endocrine active substances that have also been demonstrated to have epigenetic mechanisms of action (Gies and Soto 2013; Austrian Federal Ministry of Agriculture, Forestry, Environment and Water Management, and Environment Agency Austria 2012; European Parliament 2013). Given the specific toxicological challenges involved in the evaluation of these substances, expert circles above all have developed discourses and debates about suitable scientific and legal instruments for the responsible handling of chemicals that may not just alter the environment, but also affect human health permanently and across generations. In particular, this transgenerational dimension shifts these initially technical discussions immediately into a broader social and ethical context (Rothstein et al. 2009). The most important aspects of the risk assessment and regulation of substances with epigenetic mechanisms of action are summarised below.

3   Risk Assessment of Endocrine Active Substances

Endocrine active substances are chemicals that can influence or disrupt the normal activity of hormones. If this results in adverse effects, they are termed endocrine disrupting substances (EDCs) (WHO/UNEP 2013). EDCs can interfere with an organism's hormonal balance in three different ways (WHO/IPCS 2002):

- They mimic hormones.
- They inhibit the action of hormones (xenohormones, environmental hormones).
- They influence the synthesis, transport, metabolism and excretion of hormones.

Our hormonal system regulates growth, developmental and reproductive processes by adjusting, among other things, our metabolism, heart functioning, salt equilibrium and water balance. EDCs can have adverse effects on these processes through epigenetic mechanisms of action and cause irreparable health damage. This group of substances includes natural hormones such as oestrogen and testosterone, but also natural, plant-derived secondary substances known as phytoestrogens such as genistein. Furthermore, thousands of industrially produced chemicals, including pesticides like vinclozolin and methoxychlor, heavy metals such as chrome, arsenic and mercury, and substances that are used in plastics and household products (e.g. phthalates, Bisphenol A, alkylphenols, cf. WHO/UNEP 2013) are also classed as EDCs.

Bisphenol A (BPA) serves as a parent material for the production of polycarbonate and epoxide resins, which are used in products that consumers come into close contact with such as food packaging, as well as the manufacture of baby bottles. Over the last few years, health damaging effects caused by BPA have been demonstrated in animal experiments and epidemiological studies. The impacts on fertility and the reproductive organs are of particular significance (Newbold et al. 2009). However, there have also been reports of adverse effects on the kidneys and prostate (Nilsson et al. 2012). Furthermore, BPA's influence on brain development (Leranth et al. 2008), and its significance in the occurrence of obesity and diabetes (Porta and Lee 2012) have been confirmed. Apart from this, epidemiological investigations on humans have demonstrated a link between BPA and cardiovascular conditions (Lang et al. 2008). The studies of intergenerational effects are particularly alarming (Susiarjo et al. 2007).

Traditional risk assessment generally seeks to establish a relationship between the type of chemical, its dosage, and its biological or toxicological effects. However, the evaluation of EDCs calls this classic risk assessment methodology fundamentally into question. In contrast to how exposure has been understood hitherto, the epigenetic effects of EDCs depend on critical phases of the exposure, which therefore means they are *time-dependent*. This can be explained by the fact that development and gene expression are not controlled continuously, but usually during very specific time windows. For example, living organisms are particularly sensitive to epigenetic changes caused by environmental chemicals during the formation of new epigenetic patterns in early embryonic development. Apart from pregnancy, the sensitive time windows also include babyhood, infancy and puberty. Exposure to chemicals during these developmental phases can lead to a higher risk of specific conditions in adulthood (Schug et al. 2011). However, the food situation during these sensitive time windows can also have an effect on long-term health (cf. Bode in the present volume). In particular, scientific investigations that have examined the connection between the food consumed in infancy and impacts on the metabolism have focussed on the development of type 2 diabetes (Stöger 2008). The various ways different organs develop mean that, apart from the time-dependent effects of EDCs, effects also arise that are dependent on the specific *anatomical site of the exposure*.

Generally, it is to be taken into consideration in hazard identification for EDCs that epigenetic effects do not result from classic forms of damage, but from errors in the control of genetic processes. This hampers attempts to determine direct causality between epigenetic changes and adverse effects (Schug et al. 2011; EFSA 2013). Multiple epigenetic mechanisms can impact on gene expression synergistically, independently or antagonistically. The effects observed do not result monocausally from the mechanisms of action that precede them. The implication is that several causes may be relevant to many different effects and that cumulative risks have to be taken into consideration. This makes it very difficult to isolate any single factor and prove beyond all doubt that it is inflicting particular damage. Recently, therefore, demands have increasingly been made for a holistic, multidisciplinary perspective to be adopted as far as the risk assessment of chemicals is concerned (Gies and Soto 2013).

Apart from this, on account of the complicated dynamics of hormone receptor occupation, deviations from the classic linear dose-response curve are found in the low dose range for EDCs. This means that effects may be demonstrated at low doses, but not occur at high concentrations. Furthermore, the effects of EDCs are not restricted to the duration of the stimulation or exposure, but function like a kind of 'memory' in the later course of the organism's life or sometimes only become apparent in following generations (Schug et al. 2011).

The unique characteristics of EDCs that have been described represent a major challenge for toxicology and disprove several elements of the traditional paradigm for the biological effects of chemicals:

1. If high doses of a chemical are not harmful, this does not rule out the possibility that low doses will be damaging (which conflicts with Paracelsus's proposition that, 'The dose makes the poison').
2. The point in time when the exposure occurs determines its toxicity (*time-specific effect*).

3. The anatomical site of the exposure determines its toxicity (*tissue-specific effect*).

However, specific, standardised testing strategies that take account of both the different epigenetic modifications in various types of tissue and individuals, and the influence of the test design are still largely lacking. The risk assessment methods that are currently in use continue to be based on the assumption of a monotonic dose-response relationship. There is therefore still a considerable need for research to develop specific analytical systems adapted to the needs of risk assessment. Mention should be made here of the proposals that have been put forward for suitable screening methods that would use specific biomarkers, or the application of systems biological approaches that would exploit OMICS techniques (genomics, proteomics, metabolomics). In parallel to further research efforts, however, better dialogue should also take place at the interface between scientists and risk assessment experts (Austrian Federal Ministry of Agriculture, Forestry, Environment and Water Management, and Environment Agency Austria 2012).

The risk assessment of dietary exposure to BPA conducted by the European Food and Safety Authority (EFSA) defines the tolerable daily intake (TDI) level at 50 µg/kg per day (EFSA 2006). However, numerous authorities in European and non-European countries, including France, Sweden, Denmark, Canada and the USA, are of the opinion that this recommended level is too high. Toxicological studies have demonstrated damaging effects below the TDI level (Gies and Soto 2013). In addition to this, cumulative risks were not taken into consideration when the tolerable level was calculated. Although the EFSA has updated its scientific recommendations several times in response to such findings, the TDI level has been retained (EFSA 2010). In the meantime, a complete reevaluation of BPA has begun that is also taking account of exposure sources other than foodstuffs. Apart from this, one recent publication deals with criteria for the identification of endocrine disruptors and problems relating to suitable testing methods (EFSA 2013). In this scientific opinion, the authors emphasise that not all endocrine active substances are endocrine disruptors. Rather, their classification as such depends on whether well founded evidence is available that the substance in question can cause harmful effects on account of the way it influences the hormonal system. They come to the conclusion that the current tests for mammals and fish and to a lesser extent those for birds and amphibians as well, are generally appropriate and cover important endocrine pathways. However, no single test is enough to decide whether a particular substance is an endocrine disruptor or not. Rather, various tests have to be conducted, the results of which are to be evaluated in their entirety by experts (EFSA 2013):

[Endocrine disruptors (EDs)] can be identified according to three criteria: endocrine activity, adversity of effects and a plausible link between endocrine activity and adverse effect. [...] [E]valuation methods should, in principle, be fit for the purpose of establishing safe doses/concentrations of EDs if

(1) certain aspects (follow up of exposure in critical windows of susceptibility to later life stages, combined exposure to multiple substances, low dose effects and non-monotonic dose response curves) are addressed and
(2) used with all available information in a weight-of-evidence approach.

4 Regulation of Endocrine Active Substances

There are EDCs in many everyday goods such as plastic bottles, washing powders and liquids, toys, skincare products, textiles, electronic articles, furniture, flooring materials and pesticides (European Parliament 2013). Furthermore, medical products (e.g. on neonatal wards), drinking water pipes and thermal paper, for example, contain Bisphenol A as well. This means it is necessary to assume continuous exposure, which is critical particularly during specific phases of development such as pregnancy or infancy. The German Environmental Study for Children actually found BPA in 99 % of urine samples from children in Germany (UBA 2009). The Endocrine Society also makes the point that EDCs represent a significant potential concern for human health (Diamanti-Kandarakis et al. 2009). In the opinion of consumer and environmental organisations, this assessment legitimises preventive state action in accordance with the precautionary principle, which is the foundation for European environmental and consumer policy (Article 191(2) of the Treaty of Lisbon 2007):

Union policy on the environment [...] shall be based on the precautionary principle and on the principles that preventive action should be taken [...].

The Commission has formulated the precautionary principle in concrete terms to the extent that, particularly in cases where there are scientific uncertainties, preventive action is justified if there is cause for concern (Commission of the European Communities 2000):

Accordingly, the precautionary principle must be applied in practice particularly in cases where, based on impartial scientific evaluation, there is cause for concern that the potential hazards for the environment and for the health of people, animals or plants are not acceptable or could be irreconcilable with the high level of protection.

To date, EDCs have been dealt with on a sectoral basis along with other substances in the following product-specific regulations (cf. also Robienski in the present volume):

- Regulation concerning the Registration, Evaluation, Authorisation and Restriction of Chemicals (Regulation (EC) 1907/2006, REACH),
- Regulation on classification, labelling and packaging of substances and mixtures (Regulation (EC) 1272/2008, CLP),
- Food Contact Materials Framework Regulation (Regulation (EC) 1935/2004),
- Regulation concerning the placing of plant protection products on the market (Regulation (EC) 1107/2009),
- Biocidal Products Regulation (Regulation (EU) 528/2012) and
- Cosmetics Regulation (Regulation (EC) 1223/2009).

On closer analysis, however, it rapidly becomes clear that the provisions set out in these pieces of legislation are not tailored to the specific challenges involved in the risk assessment and responsible risk management of endocrine active substances. For instance, 89 international scientists stated in the *Berlaymont Declaration* that European chemicals law was 'entirely inadequate' for the identification of EDCs (The 2013 Berlaymont Declaration on Endocrine Disrupters 2013). They argued existing options, such as the provisions on the restriction and authorisation of substances set out in Article 57 of the REACH Regulation, ought to be exploited more comprehensively. Apart from this, the classification of BPA under the CLP Regulation as 'suspected of damaging fertility' is felt to be insufficient. BPA is permitted in food packaging by Commission Regulation (EU) 10/2011. Since 2001, however, one product-specific directive has placed restrictions on BPA, banning its use in the production of baby bottles made of polycarbonate in the EU (Commission Directive 2011/8/EU). France and Sweden have banned BPA in all food packaging for infants under stricter national rules and regulations. As of 2015, France even wishes to prohibit the use of BPA in all food packaging. The French Agency for Food, Environmental and Occupational Health and Safety (ANSES) is therefore critical of the inadequate European precautionary measures. According to recent press reports, Sweden too wants to ban the chemical completely (Svenska Dagbladet 2013).

The biggest problem with the extant European regulations is the lack of suitable, specific testing methods for the identification of EDCs and the assessment of their effects. For example, the active substance in a pesticide will only be authorised if it does not have any endocrine active properties that may cause adverse effects in humans (Regulation (EC) 1107/2009).

However, no concept for the testing of active substances to uphold this standard has been drawn up so far. There continues to be a failure to reach consensus on unambiguous regulatory definitions for the terms 'endocrine active substance' and 'endocrine disruptor'. The European Food Safety Authority (EFSA) distinguishes endocrine active substances without harmful properties from endocrine disruptors whose endocrine activity is causally connected with harmful effects. Under the new EU Biocidal Products Regulation, the Commission was supposed to specify scientific criteria that would allow biocidal products' endocrine activity to be determined by 13 December 2013. However, this deadline was not met because a public consultation was still supposed to be conducted before the criteria were finalised.

The European Parliament has therefore adopted a resolution on the protection of public health from endocrine disruptors (European Parliament 2013), which is currently the subject of negotiations between the Commission and the Council. The recitals state that hormone-related disorders and illnesses have increased in recent years. These include impaired sperm quality, increased incidence of deformed sexual organs and increased incidence of certain forms of cancer and metabolic diseases. Such a rapid increase in the incidence of these conditions can only be attributed to external environmental factors and not to genetic changes. The European Parliament refers to the 27,000 research reports that have discussed endocrine disruptors, and their effects on humans and animals to date. Among other things, the following arguments are mentioned for legislation to be adopted:

- In line with the precautionary principle, the Commission should take adequate measures to reduce human exposure to endocrine disruptors to a minimum.
- Although there is legislation that contains legal provisions concerning endocrine disruptors, there are no criteria on which to decide whether a substance should be regarded as having endocrine disrupting properties. This undermines the proper application of the existing legislation.
- The standard data requirements under EU chemicals legislation are insufficient to identify endocrine disrupting properties in an adequate manner. Endocrine disruptors should be classified as Substances of Very High Concern within the meaning of the REACH Regulation.
- A number of EU laws are aimed at protecting citizens from exposure to harmful chemicals, but each exposure is assessed individually and there is no integrated assessment of cumulative effects.
- 'Endocrine disruptors' should be introduced as a regulatory class.

- Testing methods and guidance documents should be developed so as to take better account of endocrine disruptors, possible low dose effects, combination effects and non-monotonic dose-response relationships.
- The information about endocrine disruptors provided for consumers should be reliable and presented in appropriate forms.

The resolution enumerates measures that ought to be taken rapidly in order to provide better protection, at least for vulnerable groups. In particular, mention is made of limits on the use of endocrine disruptors in goods such as skincare products, textiles and toys. Generally, it becomes clear that new regulatory projects on endocrine disruptors should take account of the challenges posed by the identification of these substances and their risk assessment in greater detail.

---

5   Summary and Outlook

Epigenetics studies the heritable information that is not contained in the genome itself. Changes to the epigenome can have negative consequences for the organism affected and also cause impacts on following generations. However, responsible action is hampered if it is not only deliberate influences, but also unintentional exposures to epigenetically active substances that have to be taken into consideration. For instance, epigenetic mechanisms of action play an important role in the harmful effects of environmental chemicals, endocrine disruptors in particular. In consequence, epigenetics represents a gateway to the understanding of a new kind of exposure with long-term impacts on human health that have been disregarded up until now. What makes the situation even more difficult is that the scientific evaluation of epigenetic mechanisms of action demands a paradigm change in classic toxicology. There will have to be detailed reflection on heuristics such as the conceptualisation of 'adverse effects'. Epigenetics therefore also poses new challenges to environmental legislation and the regulation of consumer products. What will be decisive for the legal treatment of epigenetically active substances will be the drafting of a clear regulatory definition designed to facilitate decisions about whether substances can be classified as disruptors or harmful. This definition would then have to be incorporated into the current legislation. Furthermore, the existing legislative instruments would have to be reviewed, gaps identified and possible regulatory measures used more comprehensively. Particular significance

attaches to information for consumers about the risks posed by endocrine disruptors and the options for their prevention, e.g. mandatory labelling for products
with which consumers come into close contact.

References

Austrian Federal Ministry of Agriculture, Forestry, Environment and Water Management,
   and Environment Agency Austria. (2012). Epigenetik in der Risikoabschätzung:
   ExpertInnen-Workshop: "Berücksichtigung von epigenetischen Effekten in der Risikoabschätzung".   http://www.umweltbundesamt.at/umweltsituation/gentechnik/gentechnik_
   termine/epigenetik/. Last accessed May 8, 2016.
Commission Directive 2011/8/EU of 28 January 2011 amending Directive 2002/72/EC as
   regards the restriction of use of Bisphenol A in plastic infant feeding bottles: *OJ L* 26/11,
   January 29, 2011.
Commission of the European Communities: Communication from the Commission on the
   precautionary principle, COM. (2000). 1 final, Brussels, February 2, 2000.
Commission Regulation (EU) No. 10/2011 of 14 January 2011 on plastic materials and
   articles intended to come into contact with food: *OJ L* 12/1, January 15, 2011.
Diamanti-Kandarakis, E., Bourguignon, J. P., Giudice, L. C., Hauser, R., Prins, G. S., Soto,
   A. M., et al. (2009). Endocrine-disrupting chemicals: An Endocrine Society scientific
   statement. *Endocrine Reviews, 30*(4), 293–342.
European Food Safety Authority (EFSA). (2006). Opinion of the scientific panel on food
   additives, flavourings, processing aids and materials in contact with food (AFC) related
   to 2,2-BIS(4-HYDROXYPHENYL)PROPANE. *EFSA Journal, 428*, 1–75.
European Food Safety Authority (EFSA). (2010). Scientific opinion on Bisphenol A:
   Evaluation of a study investigating its neurodevelopmental toxicity, review of recent
   scientific literature on its toxicity and advice on the Danish risk assessment of
   Bisphenol A. *EFSA Journal, 8*(9), 1829–1945.
European Food Safety Authority (EFSA). (2013). Scientific opinion on the hazard
   assessment of endocrine disruptors: Scientific criteria for identification of endocrine
   disruptors and appropriateness of existing test methods for assessing effects mediated by
   these substances on human health and the environment. *EFSA Journal, 11*(3), 3132.
European Parliament. (2013). Resolution: Protection of public health from endocrine
   disruptors, 2012/2066(INI), March 14, 2013 (P7_TA(2013)0091).
Feinberg, A. P. (2007). Phenotypic plasticity and the epigenetics of human disease. *Nature,
   447*(7143), 433–440.
German Federal Environment Agency (UBA). (2009). In K. Becker, H. Pick-Fuß, A.
   Conrad, C. Zigelski, M. Kolossa-Gehring, T. Göen, & A. Seidel (Eds.),
   *Kinder-Umwelt-Survey (KUS) 2003/06, Human-Biomonitoring-Untersuchungen auf
   Phthalat- und Phenanthrenmetabolite sowie Bisphenol A,* Funding Code 01 EH 0202.
   Umwelt und Gesundheit 4/2009.
Gies, A., & Soto, A. M. (2013). Bisphenol A: Contested science, divergent safety
   evaluations. In European Environment Agency (EEA) (Eds.), *Late lessons from early*

warnings: Science, precaution, innovation, chap. 10 (pp. 247–71). http://www.eea. europa.eu/publications/late-lessons-2. Last accessed May 8, 2016.

Jirtle, R. L., & Skinner, M. K. (2007). Environmental epigenomics and disease susceptibility. *Nature Reviews Genetics, 8*, 253–262.

Lang, I. A., Galloway, T. S., Scarlett, A., Henley, W. E., Depledge, M., Wallace, R. B., et al. (2008). Association of urinary Bisphenol A concentration with medical disorders and laboratory abnormalities in adults. *JAMA, 300*(11), 1303–1310.

Leranth, C., Hajszan, T., Szigeti-Buck, K., Bober, J., & MacLusky, N. J. (2008). Bisphenol A prevents the synaptogenic response to estradiol in hippocampus and prefrontal cortex of ovariectomized nonhuman primates. *Proceedings of the National Academy of Sciences of the United States of America, 105*(37), 14187–14191.

National Research Council (NRC). (1983). *Risk assessment in the federal government: Managing the process.* Washington, DC: The National Academies Press.

Newbold, R., Jefferson, W. N., & Padilla-Banks, E. (2009). Prenatal exposure to Bisphenol A at environmentally relevant doses adversely affects the murine female reproductive tract later in life. *Environmental Health Perspectives, 117*(6), 879–885.

Nilsson, E., Larsen, G., Manikkam, M., Guerrero-Bosagna, C., Savenkova, M. I., & Skinner, M. K. (2012). Environmentally induced epigenetic transgenerational inheritance of ovarian disease. *PLoS ONE, 7*(5), e36129.

Owen, C. M., & Segars, J. H, Jr. (2009). Imprinting disorders and assisted reproductive technology. *Seminars in Reproductive Medicine, 27*(5), 417–428.

Porta, M., & Lee, D.-H. (2012). *Review of the science linking chemical exposures to the human risk of obesity and diabetes.* Chem Trust report. http://www.chemtrust.org.uk/wp-content/uploads/CHEM-Trust-Obesity-Diabetes-Full-Report.pdf. Last accessed May 8, 2016.

Regulation (EC) 1107/2009 of the European Parliament and the Council of 21 October 2009 concerning the placing of plant protection products on the market and repealing Council Directives 79/117/EEC and 91/414/EEC, *OJ L* 309/1, November 24, 2009.

Regulation (EC) 1223/2009 of the European Parliament and the Council of 30 November 2009 on cosmetic products, *OJ L* 342/59, December 22, 2009.

Regulation (EC) 1272/2008 of the European Parliament and the Council of 16 December 2008 on classification, labelling and packaging of substances and mixtures (CLP), amending and repealing Directives 67/548/EEC and 1999/45/EC, and amending Regulation (EC) No. 1907/2006, *OJL* 353, December 31, 2008.

Regulation (EC) 1907/2006 of the European Parliament and the Council of 18 December 2006 concerning the Registration, Evaluation, Authorisation and Restriction of Chemicals (REACH), *OJ L* 396/1, December 30, 2006.

Regulation (EC) 1935/2004 of the European Parliament and the Council of 27 October 2004 on materials and articles intended to come into contact with food and repealing Directives 80/590/EEC and 89/109/EEC, *OJ L* 338/4, November 13, 2004.

Regulation (EU) 528/2012 of the European Parliament and the Council of 22 May 2012 concerning the making available on the market and use of biocidal products, *OJ L* 167/1, June 27, 2012.

Rothstein, M. A., Cai, Y., & Marchant, G. E. (2009). The ghost in our genes: Legal and ethical implications of epigenetics. *Health Matrix Clevel, 19*(1), 1–62.

Schug, T. T., Janesick, A., Blumberg, B., & Heindel, J. (2011). Endocrine disrupting chemicals and disease susceptibility. *Journal of Steroid Biochemistry and Molecular Biology, 127*, 204–215.

Skinner, M. K. (2011). Role of epigenetics in developmental biology and transgenerational inheritance. *Birth Defects Res C Embryo Today, 93*, 51–55.

Skinner, M. K., Manikkam, M., & Guerrero-Bosagna, C. (2010). Epigenetic transgenerational action of environmental factors in disease etiology. *Trends in Endocrinology and Metabolism, 21*, 214–222.

Stöger, R. (2008). The thrifty epigenotype: An acquired and heritable predisposition for obesity and diabetes? *BioEssays, 30*, 156–166.

Susiarjo, M., Hassold, T. J., Freeman, E., & Hunt, P. A. (2007). Bisphenol A exposure in utero disrupts early oogenesis in the mouse. *PLoS Genetics, 3*(1), e5.

Svenska Dagbladet. (2013). Vi tar strid i EU om förbud mot bisfenol, February 1, 2013. http://www.svd.se/opinion/brannpunkt/vi-tar-strid-i-eu-om-forbud-mot-bisfenol_7878460.svd. Last accessed May 8, 2016.

Tang, W., Levin, L., Talaska, G., Yun, C., Herbstman, Y. J., Tang, D., et al. (2012). Maternal exposure to polycyclic aromatic hydrocarbons and 5'-CpG methylation of interferon-[gamma] in cord white blood cells. *Environmental Health Perspectives, 120*(8), 1195–1200.

The 2013 Berlaymont Declaration on Endocrine Disrupters. (2013). http://www.brunel.ac.uk/__data/assets/pdf_file/0005/300200/The_Berlaymont_Declaration_on_Endocrine_Disrupters.pdf. Last accessed May 8, 2016.

Treaty of Lisbon amending the Treaty on European Union and the Treaty establishing the European Community, *OJ C* 306/01 (2007), consolidated versions of the Treaties published in *OJ C* 326/01 (2012).

Weinhold, B. (2012). More chemicals show epigenetic effects across generations. *Environmental Health Perspectives, 120*, 228.

World Health Organization International Programme on Chemical Safety (WHO/IPCS). (2002). Global assessment of the state-of-the-science of endocrine disruptors. http://www.who.int/ipcs/publications/new_issues/endocrine_disruptors/en/. Last accessed May 8, 2016.

World Health Organization and United Nations Environment Programme (WHO/UNEP). (2013). In Å. Bergman, J. J. Heindel, S. Jobling, K. A. Kidd, & R. T. Zoeller (Eds.), *State of the science of endocrine disrupting chemicals.* Geneva: WHO/UNEP. http://www.who.int/ceh/publications/endocrine/en/index.html. Last accessed May 8, 2016.

Author Biography

Jutta Jahnel Dr. is a researcher at the Institute for Technology Assessment and Systems Analysis (ITAS) at the Karlsruhe Institute of Technology (KIT). She is working in the field of risk governance of nanomaterials, focusing on methodical aspects and uncertainties of scientific risk assessment (Conceptual Questions and Challenges Associated with the Traditional Risk Assessment Paradigm for Nanomaterials, 2015) and the potential of participatory methods for policy advice (Focus group discussions inform concern assessment and support scientific policy advice for the risk governance of nanomaterials, 2012). Going beyond a traditional natural science perspective she describes an opening-up of a narrow expert-based assessment approach (Technology assessment beyond toxicology—the case of nanomaterials, 2014). Jutta has also investigated the challenges in assessing and regulating other types of new chemicals of concern such as endocrine disruptors. Her other research interests are the conceptualisation and implementation of the principles of 'Responsible Research and Innovation' especially with regard to the governance of new and emerging science and technologies (NEST).

Contact: KIT—The Research University in the Helmholtz Association, Institute for Technology Assessment and Systems Analysis (ITAS), Karlstraße 11, D-76133 Karlsruhe.

Epigenetics and Legal Regulations: A Challenge for the State's Obligation to Protect Environment, Individual's Health and Civil Rights

Jürgen Robienski

Abstract

Epigenetics is a central topic of research in the life sciences at the beginning of the twenty-first century. As we become aware of the fact that genes and the environment interact with each other in a dialectic process, the dogma of genetic determinism is proven wrong. The impact of environmental factors can damage genes as well as modify genetic functions. This may cause diseases. Epigenetics therefore contributes to a better understanding of diseases and opens up new possibilities for diagnostics and treatment. At the same time, this brings up several legal questions, including a risk assessment of new as well as of conventional technologies and noxious substances, the role of preventive public health care, the social reimbursement, responsibility, and liability for any epigenetically relevant environmental impact, and assessment based on medical legislation.

1   Introduction

The life sciences are inspired by a new spirit: the "spirit in our genes," namely epigenetics. Epigenetics belongs to the larger field of molecular genetics. There is no unequivocal definition for it as yet. It is based on recognition of the fact that "a gene is no autonomous entity, no distinct DNA sequence which always produces

J. Robienski (✉)
Eichenkamp 6, 38539 Müden, Germany
e-mail: robienski@aol.com

© Springer Fachmedien Wiesbaden GmbH 2017   141
R. Heil et al. (eds.), *Epigenetics*, Technikzukünfte, Wissenschaft und
Gesellschaft / Futures of Technology, Science and Society,
DOI 10.1007/978-3-658-14460-9_11

the same effect" (Jablonka and Lamb, cited after Shenk 2012, p. 33). Rather, genes, proteins, and environmental signals interact continuously, influencing genetic regulation (Shenk 2012, p. 47) and the external appearance of organisms. The dogma of "nature versus nurture" has been replaced by the "epigenetic dogma" of "nature and nurture" as a process of dialectic interaction between gene and environment (Shenk 2012, p. 47). As a connecting link between environmental impact and genes, epigenetics may be able to transcend genetic determinism without questioning the importance of genetics as such. Many times it is the combination or, rather, interaction between genetic makeup, environmental impact, and the epigenetic changes induced by the latter that causes diseases. The aim of epigenetic research is therefore to identify any environmental impact that evokes epigenetic changes, whether alone or in combination with other factors. Epigenetic changes may appear as modifications of genetic functions inheritable by mitosis and/or meiosis that cannot be explained by changes in the DNA sequence, such as DNA methylation and changes in the chromatin- (e.g., by modifications of histones) and RNA-mediated gene regulation mechanisms (Kegel 2011, p. 81 with reference to Felsenfeld; Youngson and Whitelaw 2008).

Since environmentally induced modifications of genetic functions may be inheritable (Kegel 2011, pp. 97, 191f.), epigenetics is described as the "substantial base for—even cross-generational—communication between environment and genome" (Spork 2009, p. 46 with reference to Albert Jeltsch).

Epigenetics research has also a scope of application called "epigenetic technological application". The aim is to identify epigenetic biomarkers that can be used for diagnostic purposes (Johnen et al. 2013, p. 20; Frank 2011, pp. 255f.) and to develop substances and techniques that make it possible to intentionally induce epigenetic modifications in order to treat diseases (Rothstein et al. 2009, p. 35; Schenk et al. 2012, p. 605). However, this area is just beginning to develop.

For a legal discussion, it is important to note that epigenetics will produce new insights into the interactive effects and risk potential of any known environmental impact (e.g., of noxious substances and other environmental stimuli) on our genome and on genetic regulation. Among the impact factors are chemicals (e.g., Bisphenol A, Nicotine, and Benzol), nutrition, and other environmental factors (e.g., shift working, traumatic events, stress, and noise). This impact may induce changes on a molecular level, which will then cause epigenetic modifications contributing to the development of diseases or an enhanced susceptibility to disease in the person concerned or their descendants.

Epigenetics has rarely been a topic of discussion in the legal sciences. There is only one more detailed essay known from the United States dealing with several legal aspects of epigenetics such as risk assessment and the questions of liability and discrimination (Rothstein et al. 2009). No comparable study can be found in Germany. However, the German Federal Environmental Agency (*Umweltbunde-samt*) recently initiated a research project which shall—among other topics—also focus on risk assessment (UBA 2013). In 2012, the Austrian Federal Environmental Agency hosted an expert workshop on dealing with epigenetic effects in risk assessment (UBA-Ö 2012). Nevertheless, in addition to environmental risk assessment, epigenetics will be relevant for all areas of legislation concerning the adverse effects of environmental factors on human health. This environmental impact must be understood in a rather broad sense here, so much so that consideration should even be given to installing a kind of epigenetic mainstreaming, i.e., to investigating the epigenetic relevance of all ecological, technical, and social projects. However, a prerequisite for this is overcoming the currently prevailing doubts about the validity of the findings of epigenetics research. Legal consequences can be drawn only if a chain of causality can be established between an environmental factor, the epigenetic modifications as its impact, and diseases or health disorders, and risks to health as a consequent of the modification with sufficient probability in a legal sense. The following discussion is therefore dependent on the validation of scientific results at a level which is high enough to serve as proof in a legal sense. The dependence on the findings of natural science illustrates the dilemma in the area of technology law.

2   Epigenetics and the State's Responsibility

The constitutional principle of the welfare state establishes the state's responsibility for society, its individuals, and its interests. As holder of the monopoly on the use of force, the state must take care to limit all risks to society. Article 20a of the German constitution (the German word is *Grundgesetz* (GG), which literally means fundamental law) establishes the obligation for the state to protect the environment and to preserve natural resources. Each individual is protected by the constitution due to the right to life and physical integrity laid down in Art. 2 II 1 GG. Article 2 II 1 urges the state to protect and promote life, to defend it against illegal attacks from others, and to preserve the individual against "all effects which may impair human health in a biological-physiological sense" (BVerfGE 54, 54/74).[1] The scope of protection established by Art. 2 II 1 GG comprises every living human as well as "nascent life," i.e., the *nasciturus* (Latin for someone who

---

[1]All translations of German legal texts, court decisions and regulations are by the author.

will be born). This term refers to a conceived but yet unborn child, i.e., the embryo after implantation in the uterus but not before implantation. Since, however, disease-causing epigenetic modifications may be triggered by environmental factors even before implantation, the scope of protection offered by Art. 2 II 1 GG should be enlarged to encompass every embryo as defined by the embryo protection law (ESchG), which has already been realized in nonconstitutional law.

Furthermore, the legal permissibility of religiously motivated physical procedures such as the circumcision of boys may need to be reconsidered, as it is suspected of triggering childhood trauma accompanied by epigenetic modifications (Olthus-Molthagen 2013). If this suspicion is verified, a new evaluation process will have to take place to decide whether the parents' right to religious freedom really takes precedence over the child's right to physical integrity.

2.1 The Precautionary Principle

Environmental and human health policies are governed by the precautionary principle. Stress and adverse effects on the environment and human health shall be prevented in advance if possible or at least minimized. The precautionary principle can justify governmental measures of risk regulation—even if the risk is not fully known—given that on the basis of scientific knowledge (risk information and evaluation) the risk is not only a theoretical, speculative one. The government has the burden of demonstration and proof. The findings of epigenetics may render the currently practiced risk assessment for various kinds of environmental impact insufficient or even question its appropriateness. In conjunction with the precautionary principle, they may even be the basis for prohibiting the further use of particular noxious substances (Rothstein et al. 2009, p. 31). As yet, there is lack of sufficient risk information as well as of methods of risk evaluation (UBA-Ö 2012, p. 7; for further information on the precautionary principle and risk assessment, see Jahnel in this issue).

2.2 Preventive Public Health Care

As far as regulatory legislation for risk management is concerned, the precautionary principle also applies to preventive public health care. This principle is applied with special stringency to occupational safety, in particular to the protection of pregnant women (maternity protection). The employment of pregnant women is prohibited if they would be exposed to health hazards (e.g., exposure to carcinogenic substances such as nicotine or benzol). In this case, it is irrelevant whether the contamination remains below official limit values or not. A harmful effect according to § 4 Abs. 1 MuSchG (maternity protection law) is even given if there is a risk that potentially harmful substances would cause damage to the woman's health.[2] In addition, every employer is obliged to carry out a risk assessment for a pregnant woman's workplace. So far, the epigenetic effects of the risks at the workplace are not considered. With regard to the epigenetic relevance of noxious substances a pregnant woman may encounter at her workplace, this must change. Even the potential risk of epigenetic effects for the pregnant woman and even more so for the unborn child should be included in the assessment.

Health promotion and prevention also belong to the field of public health protection. In the light of the epigenetic consequences of harmful nutrition (e.g., too much fat and sugar, alcohol and nicotine consumption) for the woman as well as for the unborn child (Lehnen et al. 2010; Hutterer 2011)—even for the woman planning to become pregnant—it has been postulated that the findings of epigenetics should be included in the respective prevention and health promotion programs (Henke 2014). The association of pediatricians of the state of Hamburg has initiated a specific prevention program in 2014 (Werner 2014). A draft prevention law prepared by the German Ministry of Health, then led by Minister Bahr, included legal provision for the implementation of specific prevention programs applicable from early childhood onwards.[3]

3 A Civil Right to Compensation and Social Right to Reimbursement

The findings regarding epigenetics are relevant in civil as well as social regulations about compensation (Rothstein et al. 2009, p. 43). If damages are due to the epigenetic impact of an environmental act, it must be clarified who caused the damage and if the respective person is responsible according to criminal as well as civil law, i.e., if a claim to compensation and/or a legal liability for the damages exists. If this is not the case, compensation (social security benefit) according to social rights (crime victim's compensation law, statutory accident insurance etc.) may be applicable.

---

[2]OVG Berlin decision of 13 July1992 6 S 72.92; VG Bayreuth decision of 12 July 2005, B 3 S 05/92—both juris.

[3]Draft of a law to promote prevention from 2013 (Bundesgesundheitsministerium 2013).

3.1 Civil Right to Compensation

The basic legal norm for the civil claim to compensation (law of torts) is § 823 of the German Civil Code (*Bürgerliches Gesetzbuch* = BGB). Further claims are laid down in various special laws, such as the German Product Liability Act (*Produkthaftungsgesetz* = ProdHG) and the German Medicines Act (*Arzneimittelgesetz* = AMG).

The civil right to compensation grants claims to compensation and to damages for pain and suffering against the damaging party to the person who has been injured or suffered health damages by an illegal act committed by the damaging party. Epigenetic modifications may be viewed as damages to health. The German Federal Supreme Court[4] defines health damage as follows: "Any induction of a state which deviates in a negative way from normal bodily functions; while it is irrelevant whether pain, a profound change in one's emotional state, or the outbreak of a disease occurred." Infection of another person with HIV by deliberate unprotected sexual intercourse is therefore an impairment of their health, even if the autoimmune disease caused by HIV did not break out.[5]

Epigenetic modifications also constitute "a state which deviates in a negative way from normal bodily functions" in the sense of this jurisdiction, at least if these modifications may lead with reasonable certainty to pain, profound changes in one's emotional state, or the outbreak of a disease. Accordingly, the Local Court Erfurt viewed the blowing of cigarette smoke on another person as personal injury because of the carcinogenic contents of the smoke.[6] The Regional Court Dresden even considers the massive psychological burden of being aware of living with an elevated health risk from increased exposure to contaminants (in this case asbestos which had not been removed by the landlord in time) as an impairment of one's health.[7] This jurisdiction could also be applied to cases which are not characterized by a direct epigenetic modification, but by an increased exposure to contaminants or environmental factors associated with the risk of epigenetic modifications and related diseases.

Psychological impairment leading to lasting traumatic damages may also represent a physical injury or health damage if it is medically identifiable and may

---

[4]BGH judgement of 14 June 2005, VI ZR 179/04—juris.

[5]BGH judgement of 14 June 2005 a. a. O.; BGH judgement of 14 Dec 1953, III ZR 183/52 (infection with syphilis)—beide juris.

[6]AG Erfurt, judgement of 18 Sept 2013, 910 Js 1195/13; see also Landgericht Bonn judgement of 09 Dec 2011, 25 Ns 555 Js 131/09—148/11—alle juris.

[7]LG Dresden judgement of 25 Feb 2011, 4 S 73/10—juris.

therefore constitute a claim to compensation.[8] Traumatic incidents may cause epigenetic modifications (see Klengel et al. 2013). An examination for epigenetic modifications after a traumatic experience such as an accident, sexual or physical abuse, or severe shock may be suitable to provide medical evidence for a psychological impairment of pathological significance ("illness value"). The law of torts not only protects the living human but also the embryo (*nasciturus*) if the injuring act by the damaging party was the cause for damage to the embryo. Prerequisite for this is, however, the full proof of the primary injury to mother and embryo. For consequential damages, the burden of proof is eased.[9] If the act of the third party leads to verifiable epigenetic modifications in the child's mother that are suitable to evoke a health impairment in the child as explained above, an obligation to provide compensation also exists towards the impaired child after its birth (Rothstein et al. 2009, pp. 43f.).

3.2 Medical Liability

The potential liability for physicians with regard to epigenetics should not be underestimated. The obligation for physicians to undertake further training also comprises the field of epigenetics, and any new scientifically valid findings published in leading medical journals must be promptly taken into consideration in the physicians' daily professional routine. A time span of three months may be too long to fulfill the required timeliness.[10]

In his treatment, the physician is also obliged to consider the living conditions of his patient as far as they are known to him, particularly if they can result in serious health risks for the patient or for the unborn child of a pregnant woman: "If the treating gynaecologist recognizes a developmental disorder of the unborn child due to nicotine abuse by the mother after the 33rd week of pregnancy, enhanced risk management is required in maternity care. Omitting the necessary measures represents gross malpractice in birth management and establishes a damages claim for pain and suffering (because of the brain damage occurring in the child)."[11]

Malnutrition (too much fat and sugar), stress, and diseases such as diabetes or preeclampsia suffered by the mother may induce epigenetic changes in the unborn

---

[8]AG Leverkusen judgement of 14 June 2013, 24 C 105/13—juris.

[9]OLG Hamm judgement of 25 May 1998, 32 U 198/97; OLG Celle judgement of 02 Nov 2000, 14 U 17/00—beide juris.

[10]OLG Koblenz judgement of 20 June 2012, 5 U 1450/11—juris.

[11]OLG Munich judgement of 25 Jan 2001, 24 U 170/98—juris.

child which may lead to severe health damages in the child after birth (see
Markunas et al. 2014 for nicotine). Epigenetic modifications must also be viewed
as developmental disorders. However, they may only be detected by the physician
if the epigenetic status of the unborn child is examined. Currently, this is not
common practice, and it is not clear if it is legally permissible. Therefore, the
physician is obliged to practice enhanced risk management during pregnancy and
even after a birth if he realizes that the mother has risk factors for epigenetic
modifications.

3.3 Social Right to Reimbursement

The social right to reimbursement in Germany is part of preventive public health
care. The respective legal provisions, including statutory accident insurance
according to the German Social Security Code 7 (SGB VII), the compensation for
victims of war according to the German Federal Law on War Pensions (BVG) and
the Crime Victims Compensation Law (OEG) make up a social protection system
granting a social security benefit in the form of the restoration of health (curative
treatment) and compensation (payment of a pension).[12] The social right to reimbursement also includes (financial) compensation for health damages due to
external effects, such as accidents or health impairment due to harmful emissions at
the work place (SGB VII), external effects caused by war (BVG), or the immediate
damage to victims caused by deliberate illegal attacks (OEG).

Not only the individual who is directly affected is protected, but also her fetus.
For example, § 12 SGB VII also views "the health damage of the fetus as a
consequence of an insurance case of the mother during pregnancy" as an insured
event. The fetus is equivalent to an insured person "due to the equality of the
dangerous situation arising from the natural unity of mother and child".[13] It is
sufficient if the health damage to the fetus was caused by specific impacts that are
generally suitable to induce damage to the mother's health.[14] Within the scope of
application of SGB VII there is no social security benefit if the conception of the
child occurs after the onset of an occupational disease.[15] The Crime Victims

---

[12]Federal Supreme Court (BVerfG) judgement of 22 June 1977, 1 BvL 2/74—juris.

[13]BVerfG judgement of 22 June 1977, 1 BvL 2/74; BSG judgement of 30 April 1985, 2 RU
43/84—alle juris.

[14]BVerfG of 20 May 1987, 1 BvR 762/85: BSG judgement of 24 Oct 1962, 10 RV 583/59
on BVG: nasciturus—alle juris.

[15]BVerfG of 20 May 1987, op. cit.—juris.

Compensation Law (OEG) also protects the unconceived child (*nondum conceptum*) in case the conception resulted from an act of violence.[16]

Furthermore, within the scope of application of SGB VII, a reasonable degree of likelihood is sufficient to assume causality between the harmful incident and the damage. There must be more evidence for than against a causal connection.[17] Full proof is not required.[18] Within the scope of application of BVG, a likelihood of a causal connection between a harmful incident and its results is sufficient to establish causality between the impairment and war effects (see § 1 III BVG).[19]

The effects of war that can lead to establishing claims based on the BVG include war-induced traumata and deficiency states such as malnutrition or insufficient medical treatment. Traumata are often the result of deliberate illegal attacks according to the OEG, e.g., sexual abuse and physical violence, particularly during childhood. Traumatization may evoke epigenetic modifications which may be the reason for various diseases from increased susceptibility to infections to, for example, post-traumatic stress disorder or depression (see for example Kean 2013, p. 359; Klengel et al. 2013).

The social insurance law also requires medical statements which give an etiological explanation for the health impairment. The Regional Social Court of Hessen[20] considered a causal link between impairment to the fetus (mental disability of the born child) and a shock to the insured mother during her third week of pregnancy as possible, but not reasonably likely, because no sufficient medical evidence or etiological explanation was given. This case might have been judged differently if the mother and child had been diagnosed with atypical epigenetic modifications which allowed a causal link between the shock, the modifications, and the mental disability of the child to be established.

The findings of epigenetics may soon become relevant for further developments in the jurisdiction of social reimbursement since the courts are obliged to take new medical scientific knowledge into consideration.[21]

---

[16]BSG judgement of 16 April 2002, B 9 VG 1/01 Rechtsanwalt (Inzest)—juris, BGHZ 11, 227 (Rape and infection with a sexually transmitted disease.).

[17]LSG BW judgement of 31 Jan 2007, L 2 U 918/05—juris.

[18]Hess. LSG judgement of 09 Dec 1992, L3 U 1152/86—juris.

[19]Berlin-Brandenburger LSG judgement of 19 April 2012, L 11 VE 85/09: Theorie der wesentlichen Bedingung: es spricht mehr dafür als dagegen; Thüringer LSG judgement of 26 June 2008, L 5 VU 784/05—beide juris.

[20]Hess. LSG judgement of 29 Nov 1989, L 3 UÄ 743/87—juris.

[21]BSG judgement of 14 Nov 2013, B 9 V 33/13 B—juris.

4   Child Protection

According to Art. 6 GG, parents or legal guardians generally have the right as well as the obligation to educate their children. Most important in the education of a child are his or her best interests (Art. 3 UN Convention on the Rights of the Child, § 1 III SGB VIII). As declared by Art. 6 III GG, the state only has a monitoring function (§ 1 II SGB VIII). It is only if parents or legal guardians fail that the state is entitled to interfere in their right to educate their child. In this case, the principle "help before withdrawal" applies. Withdrawal of a child is the ultimate measure the state may resort to. The Higher District Court (OLG) Frankfurt explains: "A threat to the child's best interests justifying interference by the state is only given if— under further unaffected development of the situation—the occurrence or further manifestation of damage in the form of a developmental disorder of the child is to be expected with reasonable certainty."[22] The question here is whether adverse epigenetic modifications caused by the parents' violation of their duty to provide care (e.g., malnutrition or physical and psychological violence) should be viewed as damage to or developmental disorder of the child within the meaning of the jurisdiction. Monitoring of the epigenetic status as part of preventive medical check-ups in (young) children could be a suitable indicator which—if the status deviates from the norm—could justify the assumption of neglect and, as a consequence, an intervention by the Youth Welfare Office as the relevant regulatory authority.[23]

According to § 1 Abs. 3 SGB VIII, it is also the task of Youth Welfare (as regulated in SGB VIII) to create and/or preserve positive living conditions and a children- and family-friendly environment for young people and their families. Based on this principle, the Law for the Support of Children from 10 Dec 2008 established a legal claim for every child from the age of one year on (or even earlier in particular cases) to receive care in a day care center for children starting in 2013 (see also § 24 SGB VIII). This legislation is also a contribution to the equality of men and women, to a greater compatibility of work and family, and thus to a move to increase the birth rate. However, it does not consider the most recent scientific findings of medical and epigenetic research. According to these findings, at least 22 % of the one- to three-year-olds show an unacceptably high level of cortisol, which can be traced back to the practice of "under 3 child care". Even among the three- to six-year-olds, 20 % of the boys and 10 % of the girls do

---

[22]OLG judgement of 23 Aug 2012, 4 UF 154/10—juris.
[23]AG Frankfurt, judgement of 16 Dec 2012, 457 F 6281/12 SO.

not feel comfortable in the children's day care unit and display an increased level of cortisol (Jul 2012, p. 15; Böhm 2011 with further references). Cortisol is a stress hormone with a neurotoxic effect. The WHO identifies stress as one of the most important threats to human health in the twenty-first century (Blech 2010, p. 124). Several studies indicate that stress may lead to epigenetic modifications, which may in turn induce various secondary physical and psychological diseases (e.g., a reduction in brain function, anxiety, depression, dissocial behavior, and an increased susceptibility to infections, allergies and obesity) (Behncke 2013a, b, c; Böhm 2011 with further references, 2012). If these scientific findings proved reasonably valid, the U3 (under 3) child care as currently practiced would present a health impairment supported by the state and at the same time a gross disservice to the state's future associated with high follow-up costs. The regulations on U3 day care for children would have to be reevaluated. However, the same investigations reveal that it is possible to prevent or at least minimize such adverse epigenetic effects by maintaining a high quality and adapting the U3 day care to the children's individual age or level of development (i.e., small groups, duration of care times adapted to age of the child) (Behncke 2013a, b, c; Böhm 2011 with further references, 2012). Hence, the state must provide much higher financial resources than are presently made available. This requires political will and a consensus in society.

The same is true for day care for children and young people in general. The epigenetic consequences of familial neglect, violence in the family, and malnutrition, for example, may be compensated by high quality care and support services offered by the state (Frank 2011, pp. 298f. with reference to the SERT study). These examples illustrate quite clearly that "epigenetic mainstreaming" is needed in addition to the well-known gender mainstreaming.

5   Epigenetic Technological Application

Some attempts to integrate new knowledge about epigenetics into medical diagnostics (Johnen et al. 2013) and into the treatment of diseases (Frank 2011, pp. 255f.; Rothstein et al. 2009, p. 35) have already been made.

The investigation of epigenetic modifications for medical purposes falls within the scope of the German Gene Diagnostics Act (GenDG). Epigenetic modifications are not genetic traits as defined by § 3 Abs. 4 GenDG. However, the analysis of epigenetic modifications is covered by the term "molecular genetics examination" according to § 3 Nr. 2 b GenDG (Kern 2012, p. 18). Nevertheless it is questionable whether it is appropriate to apply the strict regulations of the GenDG to analyses for epigenetic modifications (cf. Fündling in this volume). As explained in Sect. 3, it is necessary to prove causality between the damaging act and the epigenetic modification if claims for damages and compensation shall be enforced. This requires reference samples from the epigenetic status before the damaging incident. These reference samples need to be archived much longer than is currently the case because health impairments caused by epigenetic modifications are often not diagnosed for decades after the incident.

In addition, the determination of epigenetic modifications could be necessary for targeted prevention measures, particularly in the case of fetal and perinatal programming which increases the probability of such modifications (Lehnen et al. 2010).

In the context of medical treatment, the question arises how to qualify from a legal perspective a treatment specifically aimed at epigenetic modification. Is the treatment with an epigenetically effective medication to be qualified as a regular medical treatment or as gene therapy? On the one hand, the fact that no changes in the DNA sequence are made is an argument against qualification as gene therapy. On the other hand, the outcome is a specific modification of gene functions, which de facto represents a mutation (Kean 2013, p. 359). If epigenetic therapy is qualified as gene therapy, the next question is if it should be viewed as somatic gene therapy or as germ line therapy according to § 5 ESchG, which is prohibited. The fact that epigenetic modifications can be inherited by following generations— or that inheritance can at least not be excluded according to § 5 IV ESchG— indicates that it should be viewed as germ line therapy. However, no artificial modification of the genetic information, i.e., the DNA sequence, takes place. When the ESchG was adopted, its only aim was to prohibit the modification or exchange of a defect DNA sequence by intervention in gametes, pronuclei, or germ line cells since no technique was available to guarantee the integration and expression of the exchanged or repaired DNA sequence at the desired location at that time (Günther 2008, p. 245). Since there is no intervention in the germ line, it appears more appropriate to qualify epigenetic therapy as somatic gene therapy or even as "regular" medical treatment, even more so because medical treatments (including chemotherapy for cancer patients) may also induce epigenetic modifications. With the ESchG being a criminal law, the principle of legal certainty also stands against the widening of the scope of § 5 Abs. 1 and 2 ESchG. With respect to the progress of modern medicine, particularly the findings about epigenetics, it appears to be more appropriate to discuss whether the prohibition of germ line therapy in § 5 ESchG should be fully maintained.

6   Epigenetics and an Individual's Civil Rights

Our new knowledge about epigenetics presents fundamental legal challenges to society, particularly with regard to the protection of the individual freedoms guaranteed by the constitution in Germany and other countries. The literature on German constitutional law agrees that the general freedom of action comprises the freedom to contract disease and to addiction.[24] There have been an increasing number of attempts to "relativize" this freedom by rewarding health-promoting and sanctioning health-impairing behavior. In the literature, this phenomenon is discussed in the context of individualized medicine as an individual's responsibility for their health or preventive health care (Eberbach 2011; Damm 2011). Thus, § 1 I 2 Social Security Code V (SGB V = statutory health insurance) declares: "The insured persons are also responsible for their own health; they shall contribute to the prevention of disease and disability or work on overcoming their consequences by a healthy lifestyle, early participation in preventive health care measures, [...]." Health-impairing behavior is sanctioned by increased insurance payments or legal claims for recourse granted to the insurance companies. § 52 SGB V allows insurance companies to ask the insured person to pay part of the costs if the disease was caused deliberately (facultative provision), while a financial contribution by the insured person is mandatory in the case of follow-up damages after aesthetic surgery that was unnecessary from a medical point of view, as well as after piercing and tattooing. An example for a reward is the bonus system for dental treatment (§ 55 SGB V), which grants a higher subsidy for some treatments if the insured person participates in preventive examinations regularly.

It is quite conceivable that the bonus–malus system will be extended to further areas with regard to the findings about epigenetics. The failed draft for a prevention law in 2013 already included regulations for rewards and sanctions.[25] Why should it not be possible to sanction a health-impairing nutrition and lifestyle, in particular if it is continued after adverse epigenetic modifications have already been diagnosed?

There are even more relevant and very recent threats to the informal right to self-determination derived from the general right of personality (Art. 1, 2 I GG). Epigenetic research is a relatively young field of science. The large body of findings and hypotheses that has been compiled in this short time is far from being sufficiently validated. Such a validation requires long-term studies covering large

---

[24]BVerfG, NJW 2011, 2113; NJW 1967, 1795.

[25]Draft of a law on prevention (Bundesgesundheitsministerium 2013)

parts of society. Participants in these studies need to be monitored scientifically for many years, starting from birth (or even before) up to an adult age, and a huge amount of information about every participant's living conditions and health must be collected. Apart from the technical challenges arising from this (information safety management), it is unclear if it is possible to guarantee the informal right to self-determination as we know it up till now under such conditions. For example, the design of the study by the Federal Environmental Agency (*Umweltbundesamt*, UBA 2013) assumes that even small children will have to be included in such investigations. Ultimately, it is the parents who decide about their participation. The participating children thus grow up with the perception that taking part in such studies and waiving one's right not to know is "normality". In the end, they will not know the right not to know. The right to knowing and not knowing will be put at risk (see on this Fündling in this volume).

7   Conclusion

The findings about epigenetics are very promising. They can contribute considerably to progress in modern medicine and to our awareness of the organization of our living conditions. If we want to make sustainable use of these findings, many legal questions must be considered. This may lead to profound social changes.

References

Behncke, B. (2013a). *Frühkindlicher Stress in der Fremdbetreuung und seine langfristigen Folgen.*   http://www.fuerkinder.org/kinder-brauchen-bindung/experten-meinen/404-fruehkindlicher-stress-in-der-fremdbetreuung-und-seine-langfristigen-folgen?tmpl=component&print=1&layout=default, last access March 2016.
Behncke, B. (2013b). *Aktuelle Studien zu psychosozialem Stress in früher Kindheit und seine möglichen Folgen.* http://www.fuerkinder.org/files/1Psychosozialer_Stress-Behncke.pdf, last access March 2016.
Behncke, B. (2013c). *Was kommt, wenn Familie geht? - Vorbild Skandinavien?.* http://www.fuerkinder.org/kinder-brauchen-bindung/aktuelles-news/394-was-kommt-wenn-familie-geht-vorbild-skandinavien, last access March 2016.
Blech, J. (2010). *Gene sind kein Schicksal.* Frankfurt am Main: S. Fischer.
Böhm, R. (2011). Auswirkungen frühkindlicher -Gruppenbetreuung auf die Entwicklung und Gesundheit von Kindern. *Kinderärztliche Praxis, 5,* 316–321.
Böhm, R. (2012). *Die dunkle Seite der Kindheit.* http://www.fachportal-bildung-und-seelische-gesundheit.de/index.php?option=com_content&view=article&id=9:faz-artikel-4-april-2012&catid=2:uncategorised&Itemid=176, last access March 2016.

Bundesgesundheitsministerium. (2013). *Entwurf eines Gesetzes zur Förderung der Prävention.* http://www.bzaek.de/fileadmin/dl/mgr/Referentenentwurf_BMG_21012013.pdf, last access March 2016.

Damm, R. (2011). Personalisierte Medizin und Patientenrechte – Medizinische Optionen und medizinrechtliche Bewertung. *Zeitschrift für Medizinrecht, 29,* 7–17.

Eberbach, W. H. (2011). Juristische Aspekte einer individualisierten Medizin. *Medizinrecht, 29,* 757–770.

Frank, L. (2011). *Mein wundervolles Genom.* München: Hanser.

Günther, H.-L. (2008). § 5 ESchG. In H. L. Günther, J. Taupitz, & P. Kaiser (Eds.), *Embryonenschutzgesetz.* Stuttgart: Kohlhammer.

Henke, R. (2014). *Grundsatzreferat zum Thema Präventionbeim 117. Deutschen Ärztetag.* http://www.aekno.de/page.asp?pageID=11254, last access March 2016.

Hutterer, C. (2011). *Diabetes Typ 1: Vitamine für's Epigenom.* http://news.doccheck.com/de/906/diabetes-typ-1-vitamine-furs-epigenom/?author=58&context=author_detail, last access March 2016.

Johnen, G., Rozynek, P., & Brüning, Th R. (2013). Epigenetik der Biomarker. *IPA-Journal, 3,* 20–23.

Jul, J. (2012). *Wem gehören unsere Kinder.* Weinheim: Beltz.

Kean, S. (2013). *Doppelhelix hält besser.* Hamburg: Hoffmann und Campe.

Kegel, B. (2011). *Epigenetik Wie Erfahrungen vererbt werden.* Köln: Dumont.

Kern, B.-R. (2012). *Gendiagnostikgesetz.* München: C. H. Beck.

Klengel, T., Mehta, D., Anacker, C., Rex-Haffner, M., Pruessner, J. C., Pariante, C. M., et al. (2013). Allele-specific FKBP5 DNA demethylation mediates gene-childhood trauma interactions. *Nature Neuroscience, 16*(1), 33–41.

Lehnen, H., Maiwald, R., Gembruch, U., & Zechnr, U. (2010). Epigenetische Aspekte der fetalen und perinatalen Programmierung. *Frauenarzt, 51,* 542–547.

Markunas, C. A., Xu, Z., Harlid, S., Wade, P. A., Lie, R. T., Taylor, J. A., et al. (2014). Identification of DNA methylation changes in newborns related to maternal smoking during pregnancy. *Environmental Health Perspectives, 122*(10), 1147–1153.

Olthus-Molthagen, M. (2013). *Beschneidung und Epigenetik.* http://kukmomentaufnahmen. molthagen.de/2013/beschneidung-und-epigenetik.html, last access March 2016.

Rothstein, M., Cai, Y., & Marchant, G. E. (2009). The ghost in our genes: Legal and ethical implications of epigenetics. *Health Matrix Clevel, 19*(1), 1–62.

Schenk, T., Chen, W. C., et al. (2012). Inhibition of the LSD1 (KDM1A) demethylase reactivates the all-trans retinoic acid differentiation pathway in acute myeloid leukemia. *Nature Medicine, 18,* 605–611.

Shenk, D. (2012). *Das Genie in uns.* Hamburg: Hoffmann und Campe.

Spork, P. (2009). *Der zweite Code.* Reinbek: rowohlt

UBA: Umweltbundesamt. (2013). *Epigenetik Umwelt und Genom ein Zusammenspiel mit Folgen. Pressemitteilung des Umweltbundesamts.* http://www.umweltbundesamt.de/themen/gesundheit/belastung-des-menschen-ermitteln/epigenetik, last access March 2016.

UBA-Ö: Umweltbundesamt Österreich. (2012). *Zusammenfassung der Ergebnisse des ExpertInnen-Workshops: "Berücksichtigung von epigenetischen Effekten in der Risikoabschätzung".* http://www.umweltbundesamt.at/fileadmin/site/umweltthemen/gentechnik/Epigenetik-Okt2012/Zusammenfassung_Workshop_Epigenetik_30Okt2012.pdf, last access March 2016.

Werner, C. (2014). *Prävention gegen Fettleibigkeit schon im Mutterleib. Hamburger Abendblatt vom 21.07.2014.* http://www.abendblatt.de/ratgeber/wissen/article130380437/Praevention-gegen-Fettleibigkeit-schon-im-Mutterleib.html, last access March 2016.
Youngson, N., & Whitelaw, W. (2008). Transgenerational epigenetic effects. *Annual Review of Genomics and Human Genetics, 9,* 233–257.

Author Biography

Jürgen Robienski Dr. rer. publ. is a German lawyer in Hannover and Müden/Aller (Lower Saxony). He is a research fellow at the Center of Ethics and Law in the Life Sciences (CELLS) of Leibniz University in Hannover. Some of his publications include: „Die Auswirkungen von Gewebegesetz und Gendiagnostikgesetz auf die biomedizinische Forschung – Biobanken, Körpermaterialien, Gendiagnostik und Gendoping" Hamburg (2010), Verlag Dr. Kovac; „Ethische und rechtliche Aspekte im Umgang mit genetischen Zufallsbefunden, Herausforderungen und Lösungsansätze", (with Rudnik-Schöneborn, S., Langanke, M., Erdmann, P.), in: Ethik in der Medizin 2013, DOI 10.1007/s00481-013-0244-x; „Aktuelle medizinrechtliche und -ethische Herausforderungen der Pathologie" (with: Hoppe, Nils), Der Pathologe 2013 34(1). His topics of research are biomedical law, biotechnical law, labor law, biobanking, life Sciences.
Contact: Eichenkamp 6, 38539 Müden.

Epigenetics and the Protection of Personality Rights

Caroline Fündling

Abstract

For some time epigenetics has been a focus of increasing public interest. While the sequencing of the human genome was the paramount event in genetics, epigenetic research is focused on mechanisms which regulate genes and on environmental influences on gene expression. Since the appearance of the debate about genetic engineering, genetic data has been considered particularly relevant for an individual's personality because of their predictive potential coupled with their general invariability, their influence on reproductive decisions, and the significance for likewise affected relatives. Genetic data could be used to create health or personality profiles. Therefore, the individual has a "right to informational self-determination," which is part of the right of personality. This allows the individual to decide whether and to which extent third persons or institutions should be granted access to one's personal data. Because genetic information may cause personal harm, e.g., if they reveal the future onset of an incurable disease, the individual has the right to decide not to know his or her genetic status. This so-called "right not to know" is also part of the right of personality. This paper examines the influence that epigenetic information exerts on the right of personality in comparison to that of genetic information and whether a reassessment of the protection of a person's genetic data, personality, and privacy is required.

C. Fündling (✉)
Oberursel, Germany
e-mail: CaroFuendling@gmx.net

© Springer Fachmedien Wiesbaden GmbH 2017
R. Heil et al. (eds.), *Epigenetics*, Technikzukünfte, Wissenschaft und Gesellschaft / Futures of Technology, Science and Society,
DOI 10.1007/978-3-658-14460-9_12

1   Introduction

"My genes—my fate" has been the message of genetics for a long time. As part of the field of biology, genetic research is focused on the genetic makeup of organisms and the principles of passing traits from one generation to the next. In the field of human genetics, sequencing technologies have become more and more efficient, making it possible to analyze a person's genetic makeup in detail (Eberbach 2011, p. 759). Epigenetics is concerned—unlike genetics—with regulatory mechanisms that can be inherited independently of the respective gene sequence and has become an increasing focus of public interest in the last few years (see Seitz and Schuol in this volume). Headings[1] such as "Break the evil spell"[2], "Food for your genes"[3], or "Switches for the genes"[4] create the impression that the individual is no longer at the mercy of his or her genes but could even exercise a significant influence on them. The fact, that epigenetic modulations are *reversible* represents a new aspect in the discussion of the legal regulation and protection of individual personality and privacy rights in genetic research and clinical genetic testing, which hitherto concentrated on *irreversible* genetic changes (mutations).

The 2009 German Genetic Diagnosis Act (GenDG, BGBl. I 2529), in force since February 2010, provides a legal framework for genetic testing for health purposes. The common approaches to the protection of individual personality rights such as those expressed in the provisions of the GenDG must be reviewed as to whether the findings of epigenetic research cast doubt on their effectiveness and whether a renewed evaluation of certain basic assumptions is required. The following section therefore examines if the GenDG and thus the targeted protection of fundamental rights also covers the field of epigenetics. It focuses on genetic testing for health purposes and in the context of employment or insurance relationships. Furthermore, the question is raised whether the study of epigenetic effects as an aspect of extensive cohort studies has an impact on the personal rights of the research subjects. For this purpose, the concept of epigenetics and its demarcation from the "classical" genetics is illustrated (Sect. 2). Subsequently, the impact of epigenetic effects on the pathogenesis of diseases (Sect. 3) and the legal framework of genetic testing (Sect. 4) as well as of genetic research (Sect. 5) is described.

---

[1]All German newspaper headings are translated by the author.

[2]Der Spiegel, 32/2008, pp. 110ff.

[3]SZ-online of 17 May 2010. http://www.sueddeutsche.de/wissen/ernaehrung-essen-fuer-das-erbgut-1.297390. Accessed: 8 May 2016.

[4]FR-online of 31 March 2009. http://www.fr-online.de/wissenschaft/epigenetik-schalter-an-den-genen,1472788,3215270.html. Accessed: 8 May 2016.

Finally, the influence of epigenetics on the legal framework and on concepts for protecting an individual's personality and privacy is discussed (Sect. 6).

2   Epigenetics: Definition and Delimitation

First, it has to be clarified what is meant by the term "epigenetics," which was coined in 1942 by the biologist Conrad Hal Waddington (see Schuol in this volume) and how epigenetics can be delineated from "classical" genetics, in particular human genetics.

2.1   Epigenetics: Definition

The term "epigenetics" contains the word "genetics". Both fields of research have a common basis but a different focus: epigenetics is that part of the field of genetics that deals with external influences on the activity pattern of genes as well as with the dissemination of this pattern to the next generation (transgenerational effects) (see Walter in this volume). Genes are certain sequences of the DNA (deoxyribonucleic acid), together comprising the human genome, which provide the information relevant for the development of certain characteristics of an individual. The specific sequences of the genes set the code for the sequence of amino acids in the gene product, the protein, which influences the appearance (phenotype) of an organism. Whether a gene is transcribed and results in the expression of certain characteristics of the individual is a question of gene expression, which depends on many different influences, including epigenetic effects (see Walter in this volume; Stephens et al. 2013, pp. 373ff.). The regulation of gene expression could be described in nontechnical terms as "switching genes on and off". The gene sequence itself is not altered. If you were to compare the human body with a personal computer, the genetic code might be seen as some kind of "hardware", and the epigenetic code (the sum of epigenetic modifications) as a kind of "software" (Stadler 2012, p. 208). Noteworthy in epigenetics is thus that gene activity that is affected by certain environmental factors and that related influences are in principle reversible. Epigenetics could be defined as *"the inheritance of stable alterations of gene regulation and gene expression that are not based on variations of the DNA sequence itself"* (Murken et al. 2011, p. 42).

2.2   Epigenetics Compared to the Traditional Understanding of Genetics

Genetics has traditionally focused on the detection of genomic sequences and their alterations, e.g., mutations in the DNA (Szyf 2009, p. 7). These mutations can affect the whole human genome by altering the total number of chromosomes, individual chromosomes by altering their structure, or specific genes by altering their sequence. As a consequence of mutations, gene products may be missing, restricted, or functionally modified. One focus of biomedical research is on the question of if and to what extent genetic mutations may cause certain diseases or have an impact on the development of diseases. Mutations are, in contrast to epigenetic effects, not reversible (ibid., p. 8). In contrast, epigenetic changes in gene expression may—as the sequence itself is not altered—possibly be reversed by pharmacological influence or even deliberately controlled, for example, by substances that act on the methylation (ibid., pp. 11ff.). A new perspective of epigenetics is that—explored so far mainly in animal experiments—diet or social interaction can influence the epigenome, which puts more of these aspects in the foreground (ibid., pp. 8ff.). "Classical" genetics mainly deals with irreversible mutations in the genetic material, while epigenetics mainly focuses on reversible changes in gene expression and the related importance of environmental influences. In addition, the heritability of epigenetic effects and certain regulatory mechanisms are investigated. However, in the following, the focus is on the potential reversibility of epigenetic imprints.

3   Epigenetic Mechanisms and Pathogenesis

Studies, such as those of the effects of the Dutch hunger winter of 1944/45 (Roseboom et al. 2000; Veenendaal et al. 2013) or of the food supply in the North Swedish village Överkalix on the affected people and their descendants (Bygren et al. 2001) as well as animal studies (Dolinoy et al. 2006), suggest that epigenetic effects could be of importance in the development of certain diseases. Thus, the nutritional status of a mother during pregnancy or—as the results from Överkalix suggest—of male ancestors at certain developmental periods could increase the risk for diabetes, cardiovascular disease, or breast cancer (Bygren et al. 2001, pp. 53ff.). The research results of epigenetics suggest that there are certain stages of development in humans (e.g., prenatal phase or shortly before puberty) when environmental conditions have a greater impact than in other phases of life

(Rothstein 2013, p. 298). Epigenetic effects may also affect brain function. If certain genes are methylated with increasing age and thus switched to "mute," and others are demethylated and thus "active", this could affect memory function and lead to diseases such as Alzheimer's disease (Mastroeni et al. 2011, pp. 1168ff.). Therefore epigenetic influences on brain metabolism could explain the occurrence of psychosis in children whose mothers have experienced traumatic events such as war or famine (Bohacek et al. 2013, p. 316). Animal studies with rats show that the amount of maternal affection in brood care and nutrition can influence stress tolerance (Youngson and Whitelaw 2008, pp. 236ff.; Francis et al. 1999, pp. 1155ff.). Individuals who have received little attention from their mothers are characterized by a lower tolerance to stress in later life (ibid., pp. 236ff., pp. 1156ff.). It has also been suggested that epigenetic imprints can promote drug addiction (Szyf 2009, p. 10). Furthermore, chemicals from the environment (e.g., biocides) and commodities (e.g., BPA, see Jahnel in this volume), but also specific stimulants (e.g., nicotine) affect the methylation and demethylation of genes. This could also affect genes favoring tumor development (Liu et al. 2007, pp. 5900ff.). Epigenetics could therefore explain the cause of many adverse health effects, especially the so-called common diseases such as diabetes and cardiovascular disease.

4   Legal Framework of Genetic Diagnostics

The handling of sensitive personal information such as genetic data affects the—often conflicting—interests of different actors. First, handling genetic data affects the rights of the person tested. Genetic data may also reveal information about a person's relatives. Both the individual and his family could be interested either in having genetic information disclosed or in refusing to have it disclosed. The task of providing information falls to the physician. Genetic information may be of interest to a third party such as an employer or insurer, especially when it allows making predictions about the future health of the (potential) contractor. The legislature has to balance these conflicting interests in having access to genetic data on the one hand and the wish to refuse to know this information on the other hand. The legislature may fulfill this duty by establishing legal provisions which protect the fundamental rights of the citizens and weigh the conflicting rights (Dreier 2013, pp. 88f.). In the following the constitutional relevance of genetic diagnostics in the doctor–patient relationship and related regulations in GenDG are described (1), as is the range of genetic testing in the workplace and in the insurance sector, which are also regulated in the GenDG from a constitutional and a subconstitutional perspective (2).

4.1   Medical Treatment: Doctor-Patient Relationship

Since most the individual's personality and privacy are affected by genetic testing, such testing has a strong impact on fundamental rights and therefore implies that the legislature has a protective duty to provide provisions such as those in the GenDG.

Influence on Fundamental Rights: Right of Personality

On a constitutional level in Germany, the protection of personality is guaranteed by the so-called general right of personality. This fundamental right was "created" by the case law of the German Federal Constitutional Court as a conjunction of Art. 2 para. 1 (general freedom of action) and Art. 1 para. 1 (human dignity) of the German Constitution to provide comprehensive protection of an individual's personality, self-determination and privacy. It should guarantee a minimum defense against intervention by public authorities (Dreier 2013, p. 375). Thus, the more intimate particular information is classified, the stronger the protection of the privacy and the higher the requirements for the state to justify any intervention (ibid., p. 393). The general right of personality includes protection of personal data. Genetic data provide information about the present and in some cases also about the future status of health, ethnic origin, ancestry, sex, or biological relatives. It is therefore particularly sensitive information, which has to be protected. The purpose of data protection is to encrypt and control the use of personal information (Simitis 1994, p. 107). Data protection has become paramount since the advent of modern data processing. In 1983 the German government proposed a census, which led to a debate over the handling of personal data. The Federal Constitutional Court stated in its so-called "Census judgment" (BVerfGE 65, 1) that "informational self-determination" is part of the general right of personality, which enables the individual to decide for himself or herself about the disclosure and use of their personal data without state interference. The court explicitly pointed out the threat to personal rights posed by data processing and the possibilities of interlinkage in evolving information technology. Informational self-determination also played an important role in the discussion of the use of a "genetic fingerprint" to identify offenders based on the analysis of body cells. The fear that was repeatedly expressed was that the collection of genetic data could serve to create "comprehensive personality profiles" ("glass human beings") by the state (Simitis 1994, pp. 110).

With regard to an individual's self-determination, the so-called "right not to know" has to be taken into account. It is defined as "the right not to know one's genetic status" (Bund-Länder-Arbeitsgruppe 1990, p. 2). It was initially postulated in the ethical debate and starts from the premise that genetic information—due to its fundamental immutability, its predictive nature, and its impact on an individual's decision to have children – has the potential to be "life changing" (Taupitz 1998, p. 594; Wellbrock 2003, p. 78) and to cause personal harm (Retzko 2006, pp. 121ff.). It may have a significant impact on the individual's personality. Thus, the extent to which this information should be disclosed should be at one's own discretion. This also includes protection against any indirect compulsion to acquire knowledge about one's genetic makeup, e.g., because of a fear of discrimination which could lead to disadvantages in professional life or when purchasing an insurance policy (Kern 2012, pp. 4ff.). The right not to know is widely recognized in jurisprudence and in the jurisdiction of the Federal High Court (German Federal High Court, judgment of 20 May 2014, VI ZR 381/13) and has found its way into the GenDG. The sensitive information that can be produced in genetic studies is therefore of great relevance to one's personality and is subject to the comprehensive protection offered by the general right of personality.

Provisions of the GenDG

GenDG was adopted by a parliament that explicitly based its assumptions on "genetic exceptionalism," that is, on the specific nature of genetic data, since it possesses predictive power for long periods of time, is relevant to one's identity, and could potentially disclose information about relatives (Deutscher Bundestag 2008, p. 1). The aim of the GenDG is to protect one's right to privacy, one's data, and oneself against discrimination. The GenDG contains a comprehensive settlement for post- and prenatal genetic testing of people for medical purposes. The GenDG defines genetic tests as genetic analyses that are made at the chromosome level, at the molecular level (molecular structure of the DNA or RNA), or with regard to gene products. Genetic analysis must aim to detect genetic characteristics. Genetic characteristics according to the GenDG are defined as genetic information inherited or acquired between fertilization and birth. As shown in the explanatory memorandum, somatic genetic changes (i.e., changes that do not affect the reproductive cells) are therefore excluded from the scope of the GenDG (Deutscher Bundestag 2008, p. 21). This is justified by the fact that somatic genetic changes do not possess predictive power with lifetime validity, which is just one of the basic assumptions about the specific nature of genetic data and thus the basis of the GenDG (ibid., p. 21). Furthermore, the GenDG differentiates between diagnostic and predictive tests and in this respect gives requirements of varying strictness. The protection of personal rights, in particular the right to self-determination, is ensured by a mandatory reservation of a physician, the provision of comprehensive patient information that must contain an explicit reference to the right not to know, and the need for express written consent. In addition, genetic counseling, for the case that a diagnostic genetic test is necessary, has to be offered after the test has been carried out and, in the case of a predictive genetic test, should be obligatory before and after the test. The protection of the right to informational self-determination is guaranteed by the possibility of a withdrawal of consent, the prohibition of disclosure of the information without the consent of the person concerned, and regulations concerning the retention and destruction of genetic samples and data.

4.2   Employment and Insurance Relationships

The spheres of working life and (voluntary) insurance were considered as being particularly relevant in the discussion of the applicability of genetic analysis and the analysis of the data obtained by this means since such health-related information could help to improve risk calculation and therefore save costs. The primarily economic interests of employers and insurers face the individual rights of (potential) employers and policyholders.

Influence on Fundamental Rights

As an expression of the freedom of action, protected by Art. 2 para. 1 GG, employers and insurers are both free to decide for themselves whom they choose as their contractor and how to make the related selection (Kern 2012, pp. 220, 258). Genetic tests could serve as a means for selecting the most appropriate employee or the insured person with the lowest risk of disease and would thus promote discrimination (ibid., p. 220). In contrast to such cost savings, however, are the right of informational self-determination and the right not to know of employees and policyholders. Including genetic tests in examinations for or during the employment relationship or in the context of risk assessments of (voluntary) insurance could force the person concerned to disclose sensitive information to the (potential) contractor, as otherwise the contract would not be agreed on (ibid., pp. 221, 258). Especially in this very personal space, however, informational self-determination should be given priority, particularly because genetic information could potentially allow statements to be made about relatives (ibid.). If the person concerned does not know his or her genetic constitution, the (de facto) coercion to undergo a genetic examination can also affect the person's right not to know (ibid.).

Provisions of the GenDG

In the GenDG, parliament has addressed the above concerns and regarded the privacy rights of employees and policyholders as superior to the interest of employers and insurers with regard to the use of genetic analysis (Kern 2012, pp. 220, 258). In order to protect an individual's rights to informational self-determination and not to know, neither the employer nor the insurer is allowed to require genetic tests to be conducted or the test results of previous investigations to be provided, to accept such results, or to use them. The existing exceptions are very restricted. For certain types of insurance (life, disability, and long-term care) the results of previous genetic tests may be required above a fixed insured sum (§ 18 para. 1 p. 2 GenDG). Exceptions in employment relationships concern the analysis of genetic products, which are allowed as part of occupational medical examinations that detect genetic traits that could lead to an employee's serious illness in connection with a specific job (for example, by working with chemicals). Upon authorization, further diagnostic genetic tests may be permitted in connection with occupational safety and health.

5   Legal Framework for Genetic Research

The use of genetic data in the field of research is expressly excluded from the scope of the GenDG. Parliament justified this by saying that research serves a "general exploration of causal factors of human traits" which does not aim at an individual person (Deutscher Bundestag 2008, p. 20). However, research is also of relevance to an individual's personality and affects the fundamental rights of research subjects and their families. According to Art. 5 para. 3 GG, science and research are free, which primarily means freedom from state interference (Deutsch and Spickhoff 2014, p. 851). Restrictions may apply on the basis of other constitutional principles such as the research subject's right to privacy (ibid., p. 851; Dreier 2013, p. 815). Thus, self-determination must be preserved by a patient's consent after being informed about the clinical trial itself and any risks it might pose (Deutsch and Spickhoff 2014, p. 852). German law does not expressly regulate research with the exception of the clinical trials of medicinal products (ibid., p. 859). At the international level, the Declaration of Helsinki of the World Medical Association, first adopted in 1964, is recognized as a (ethical) directive regarding research on humans. Since it deals with sensitive data, the right to informational self-determination is affected. It is ensured by the relevant German data protections laws. Personal health data is explicitly included within the scope of the Data Protection Act. The Declaration of Helsinki emphasizes that the privacy of research subjects and the confidentiality of their personal information have to be protected. The use of genetic data or human body material in clinical trials raises the issue of its storage. This is done in so-called biobanks, which themselves are not yet regulated by law. The use of biobanks demonstrates the limits of data protection concepts since a genetic sample implies the risk of the reidentification of the data's subject and is associated with a wide range of very sensitive data (Rothstein 2013, pp. 299; Wellbrock 2003, pp. 78).

6 Specific Impact of Epigenetics

As a glance at the legal regulation of genetic diagnosis and research shows, personality rights play an important role but must be balanced with the conflicting interests of third parties, which must be brought into balance. The following sections provide details on whether epigenetics has an impact on existing regulations and whether a reevaluation of the basic legal protection of personality is required.

6.1 Medical Treatment

The question has to be addressed whether epigenetics ever falls within the scope of the GenDG or just that of the general medical law with its less stringent requirements. Studies of epigenetic changes, such as the methylation state of DNA, expressly fall within the scope of the GenDG (Deutscher Bundestag 2008, p. 21). It is unclear, however, whether it is a genetic characteristic in the legal sense. Although some epigenetic changes appear to be hereditary, their reversibility seems problematic. The legislature justifies its definition with the lifelong predictive power of genetic characteristics (ibid., p. 21). This would not always be given due to the variability of different epigenetic changes. Thus a genetic analysis to detect epigenetic changes would not or only partially fall within the scope of GenDG because its definition as a genetic characteristic as defined by the GenDG would be questionable. This would create considerable legal uncertainty for the doctors concerned. Thus, the legislature has to clarify the applicability of the GenDG. Furthermore, it is questionable whether epigenetic imprints have the same relevance for personality as genetic characteristics do, i.e., any changes to DNA itself. The right not to know, as a basic assumption, assumes that the immutability, i.e., the fateful nature, of certain genetic information might be perceived as a burden, such as might be posed for example by the early diagnosis of a fatal genetic disease or of a predisposition for an inherited disease that could lead the individual to abandon plans for having children. Epigenetic changes, however, are dynamic and can be reversible. Their knowledge is thus not as equally distressing as the knowledge of a hereditary disease and may even be positive in the sense of identifying a means to possibly influence a problematic state. But knowledge of the methylation status of specific genes and an associated increased risk of cancer can also be stressful as long as it is uncertain whether the risk can be influenced or not. In this regard, such knowledge could be compared with knowledge about a genetic predisposition to multifactorial diseases, which is also subject to uncertainty about the onset of the disease. In both cases, the individual retains the possibilities of seeking prevention or early detection. Due to the reversibility of epigenetic effects, however, there is in principle a larger spectrum for treatment options. The right not to know, while recognized in principle also with regard to epigenetic information, is however weakened by the possible reversibility of epigenetic effects.

6.2 Employment and Insurance Relationships

This raises the question of whether the restrictive regulations applied to genetic tests will also be applied to tests for epigenetic effects. Both in employment relationships and for signing insurance contracts, the ban on the use of genetic data is aimed to protect the right of informational self-determination and should serve to prevent the creation of health and personality profiles. Though the principal feature of immutability is lacking in epigenetic imprints, many uncertainties still remain. The detection of methylation patterns and the possible association with lifestyle data imply an even stronger risk of profiling and discrimination than does a genetic test. Epigenetic tests should therefore be ruled out of the context of reaching an agreement about employment or an insurance contract. In contrast, in some occupations epigenetic tests could be beneficial for the employee's health or safety; the findings of epigenetics could provide a new base such as for stress reduction in employment relationships. An analysis conducted for the benefit of employees should however continue to be constrained and fall under strict privacy regulations.

6.3 Research

Epigenetic effects play a role in many different areas, such as the influence of the diet of the parents and grandparents on the metabolism of offspring, the consequences of exposure to chemicals, and the effects of stress, emotional support, and the development of addictions and diseases on our understanding of so-called common diseases, such as diabetes, cardiovascular disease, and cancer. Epigenetics is therefore increasingly a part of the epidemiological studies that deal with the relationship between genes, environmental factors, and lifestyle in disease pathogenesis (Wellbrock 2003, p. 80). A concept of an eco-epidemiological cohort study—created on behalf of the Federal Environment Agency (Umweltbundesamt—UBA) by the Institute of Medical Computer Science, Biometry and Epidemiology at the University Hospital Essen—explicitly includes the study of epigenetic processes and the development of epigenetic profiles (UBA 2013, pp. 164f.). Extensive tissue samples are already taken during childhood as are also data on, among other things, pollutant exposures, genetic factors, and social factors such as family background, stress, and geographical origin (ibid., pp. 67ff.). Due to the (presumed) transgenerational effects, the importance of environmental influence and their different effects at different stages of a person's life, the study of epigenetic effects requires an extensive collection of data and their storage for long periods of time. The result is a situation of risk for the informational self-determination of the persons affected. The storage of tissue samples in biobanks and the simultaneous survey of lifestyle data increase the risk of reidentification and the creation of extensive personality profiles (Wellbrock 2003, p. 80). This once again brings the original intention of informational self-determination, namely to protect a citizen from being made "transparent," to the forefront. It must therefore be carefully monitored whether gradual erosion of this protection takes place as a result of large-scale data collection—even if it serves the laudable purpose of exploring the causes of diseases—and whether new measures for protecting data are necessary. While this is not a consequence of epigenetics in particular, but a general consequence of epidemiological studies with a large number of participants, it is enhanced by the design of studies of epigenetic processes, requiring the collection, analysis, and long-term storage of a variety of sensitive data.

7 Conclusion

Although epigenetic research is still in its infancy and many of it promises may not be fulfilled until the distant future, it turns out that the findings of epigenetic research are of relevance to the individual's right of personality and privacy. Under current legislation, the strict regulations of the GenDG concerning privacy protection only partly apply to the testing of epigenetic effects in the doctor–patient relationship. In this regard, there should be clarification by the legislature, for example, through a clear definition of epigenetics in the GenDG. Due to the—in principle—reversibility of epigenetic effects, the particular significance of the right not to know is de facto attenuated because it is based on the assumption of the immutability of genetic dispositions. The right to informational self-determination is also involved in employment and insurance relationships as well as in the field of research. Because of the link between lifestyle data and biological data, which include epigenetic imprints, the risk of creating health and personality profiles again becomes crucial. Public authorities should particularly monitor the development of the research sphere in order to counter a threat to informational self-determination at an early stage. The existing data protection concepts consequently have to be scrutinized. The positive aspects of the findings of epigenetic research such as providing new insights into the prevention or cure of diseases, such as diabetes or cancer, or even reducing stress have to be weighed against the protection of informational self-determination.

References

Bohacek, J., Gapp, K., Saab, B. J., & Mansuy, I. M. (2013). Transgenerational epigenetic effects on brain functions. *Biological Psychiatry, 73*, 313–320.

Bund-Länder Arbeitsgruppe "Genomanalyse". (1990). *Abschlussbericht, Bundesanzeiger, Nr. 161a, 29 August 1990*. Köln: Bundesanzeiger.

Bygren, L. O., Kaati, G., & Edvinsson, S. (2001). Longevity determined by paternal ancestors' nutrition during their slow growth period. *Acta Biotheoretica, 49*, 53–59.

Deutscher Bundestag. (2008). *Gesetzesentwurf der Bundesregierung, Entwurf eines Gesetzes über genetische Untersuchungen bei Menschen (Gendiagnostikgesetz – GenDG). BT-Drucksache 16/10532*.

Deutsch, E., & Spickhoff, A. (2014). *Medizinrecht* (7th ed.). Berlin: Springer.

Dolinoy, D. C., Weidman, J. R., Waterland, R. A., & Jirtle, R. L. (2006). Maternal genistein alters coat color and protects A^{vy} mouse offspring from obesity by modifying the fetal epigenome. *Environmental Health Perspectives, 114*, 567–572.

Dreier, H. (2013). *Grundgesetz Kommentar, Bd. I, Präambel, Artikel 1 – 19*, 3rd edn. Tübingen: Mohr Siebeck.

Eberbach, W. H. (2011). Juristische Aspekte einer individualisierten Medizin. *Medizinrecht, 29*, 757–770.

Francis, D., Diorio, J., Liu, D., & Meaney, M. J. (1999). Non-genomic transmission across generations of maternal behavior in the rat. *Science, 286*, 1155–1158.

Kern, B.-R. (Ed.). (2012). *Gendiagnostikgesetz, Kommentar*. Munich: C.H. Beck.

Liu, H., Zhou, Y., Boggs, S. E., Belinsky, S. A., & Liu, J. (2007). Cigarette smoke induces demethylation of prometastatic oncogene synuclein-c in lung cancer cells by downregulation of DNMT3B. *Oncogene, 26*, 5900–5910.

Mastroeni, D., Grover, A., Delvaux, E., Whiteside, C., Coleman, P. D., & Rogers, J. (2011). Epigenetic mechanisms in Alzheimer's disease. *Neurobiology of Aging, 32*, 1161–1180.

Murken, J., Grimm, T., Holinski-Feder, E., & Zerres, K. (2011). *Humangenetik* (8th ed.). Stuttgart: Georg Thieme Verlag.

Retzko, K. (2006). *Prädiktive Medizin versus ein (Grund-)Recht auf Nichtwissen*. Aachen: Shaker Verlag.

Roseboom, T. J., van der Meulen, J. H. P., Osmond, C., Barker, D. J. P., Ravelli, A. C. J., Schroeder-Tanka, J. M., et al. (2000). Coronary heart disease after prenatal exposure to the Dutch famine, 1944–45. *Heart, 84*, 595–598.

Rothstein, M. A. (2013). Legal and ethical implications of epigenetics. In R. L. Jirtle & F. L. Tyson (Eds.), *Environmental epigenomics in health and disease* (pp. 297–308). Berlin: Springer.

Simitis, S. (1994). Allgemeine Aspekte des Schutzes genetischer Daten. In: Schweizerisches Institut für Rechtsvergleichung (Eds.) *Genanalyse und Persönlichkeitsschutz, Internationales Kolloquium, Lausanne, 14 April 1994* (pp. 107–127). Zürich: Schulthess, Polygraphischer Verlag.

Stadler, S. C. (2012). Beitrag auf der 9. Jahrestagung der Deutschen Vereinten Gesellschaft für Klinische Chemie und Laboratoriumsmedizin (DGKL) 26-29 September 2012 in Mannheim, Diagnostische und therapeutische Möglichkeiten der Epigenetik in der Onkologie. *Versicherungsmedizin, 2012*, 208–209.

Stephens, K. E., Miaskowski, C. A., Levine, J. D., Pullinger, C. R., & Aouizera, B. E. (2013). Epigenetic regulation and measurement of epigenetic changes. *Biological Research for Nursing, 15*, 373–381.

Szyf, M. (2009). Dynamisches Epigenom als Vermittler zwischen Umwelt und Genom. *Medizinische Genetik, 21*, 7–13.

Taupitz, J. (1998). Das Recht auf Nichtwissen. In: Hanau et al. (Eds.), *Festschrift für Günther Wiese zum 70. Geburtstag* (pp. 583–602). Neuwied u.a.: Luchterhand.

UBA: Umweltbundesamt. (2013). *Konzept für eine umweltepidemiologische Geburtskohorte als Beitrag zur gesundheitsbezogenen Umweltbeobachtung des Bundes (GUB)*. http://www.umweltbundesamt.de/sites/default/files/medien/378/publikationen/sonstige_konzept_fffu_eine_umweltepidemiologische_geburtskohorte.pdf. Accessed 22 April 2014.

Veenendaal, M., Painter, R., de Rooij, S., Bossuyt, P., van der Post, J., Gluckman, P., et al. (2013). Transgenerational effects of prenatal exposure to the 1944–45 Dutch famine. *BJOG: An International Journal of Obstetrics and Gynaecology, 120*, 548–554.

Wellbrock, R. (2003). Datenschutzrechtliche Aspekte des Aufbaus von Biobanken für Forschungszwecke. *Medizinrecht, 21*, 77–82.

Youngson, N. A., & Whitelaw, E. (2008). Transgenerational epigenetic effects. *Annual Review of Genomics and Human Genetics, 9*, 233–257.

Author Biography

Caroline Fündling Dr. iur. is a lawyer and recently finished her Ph.D. studies at the University of Augsburg. In her thesis she focuses on legal issues of genetic diagnosis. She is interested in civil law, especially in medical law and labor law. She is currently working at a law firm in Frankfurt.

Adam's Apple and His Legacy: Ethical Perspectives on Epigenetics with an Excursion to the Field of Body Weight Regulation

Jens Ried

Abstract

Substantial potential is ascribed to epigenetics to solve previously pending questions, especially concerning the emergence and transmission of diseases. Epigenetic approaches are of particular interest for obesity research since genes associated with obesity could possibly be modulated by appropriate nutritional profiles and these alterations could then remain stable over several generations. Opportunities of more intensified transdisciplinary cooperation are arising within the field of epigenetics by the synopsis of ecological, behavioral, genetic and social factors. However, this synopsis at once raises issues, for example, concerning the justice and equality in health care or concerning the relation between genetic findings and personal responsibility.

1 Introduction

"Lifetime memory and tradition beyond individuals and generations is, biologically seen, the only form known to us of the 'inheritance of acquired characteristics.'[1] (Blumenberg 2006, p. 583). In *Beschreibung des Menschen*, Hans

---

[1]"Lebenszeitgedächtnis und Tradition über Individuen und Generationen hinweg ist, biologisch betrachtet, die einzige uns bekannte Form der 'Vererbung erworbener Eigenschaften.'"

J. Ried (✉)
Center for Management, Technology and Society, Nuremberg Campus of Technology, Fürther Str. 246c, D-90429 Nürnberg, Germany
e-mail: jens.ried@fau.de

© Springer Fachmedien Wiesbaden GmbH 2017   171
R. Heil et al. (eds.), *Epigenetics*, Technikzukünfte, Wissenschaft und Gesellschaft / Futures of Technology, Science and Society,
DOI 10.1007/978-3-658-14460-9_13

Blumenberg uses this brief sentence and thus coins an axiom that has long been standing under the hegemony of the Darwinian theory of evolution—or at least of a certain understanding of it—not only valid for biology but also for philosophical anthropology, which is sensitive for natural science: Evolutionary mechanisms are based on arbitrary mutations of the DNA (deoxyribonucleic acid) which lead to useful but random adaptations to the respectively prevailing and permanently changing ecological conditions—or to maladaptations which lower the reproduction rate and either force a species to change their ecological niche or to become extinct. According to this evolutionary dogma, all characteristics which are acquired by behavior, however, are principally excluded from being inherited. Nothing that reaches the body from outside or enters it can result in the process of inheritance unless it is mutagenic in such a sense that it affects the DNA sequence in a modulating way. Other concepts for which the heritability of acquired characteristics is at least conceivable and which are associated with the name Lamarck by those interested in the history of science have been considered to be as more or less scientifically odd for a century and a half.

With the advent of epigenetics as one of the current spearheads of the development of emerging biotechnologies and life science innovations (Berdasco and Esteller 2013), a small late victory for Lamarck, or at least the temptation to accept it (Haig 2007), seems now to emerge. The novelty value which is attributed to this epistemological development also described as "epigenetic turn" (Nicolosi and Ruivenkamp 2012) essentially depends on how the paradigm, set as overcome or crucially modified, is defined. Therefore, it is no wonder that the circulated and discussed assumptions of epigenetics reach from a paradigm shift to a hardly surprising, because expectable, next step to genomics.

The controversies found rather on the level of theory produce an interpretation which is expressed explicitly or assumed implicitly. This interpretation does not only guide the scientific, but also the cultural and sociological and not least the ethical perspective. Therefore, it is not only possible from an ethical perspective, but inevitable to map the diffuse field of epigenetics to be able to appropriately decide on the novelty and the intensity of the evoked ethical conflict situations. Due to the wide extent of this field, it can only be done paradigmatically at this point. Since the various theoretical readings and interpretations have to prove themselves not least on the level of application, the following deliberations will concentrate on obesity as an essential risk factor for almost the whole range of *non-communicable diseases*,[2] to which nowadays the biggest health problems are linked, not only in the western world (Bauer et al. 2014), and on which the main focus of the WHO is (WHO 2013). Furthermore, these deliberations will outline it as a field of research which does not accidentally attract epigenetic models and by means of which the practical relevance of epigenetic findings can be examined (Conn et al. 2013; van Vliet-Ostaptchouk et al. 2012; Lavebratt et al. 2011; Gluckman and Hanson 2008). Therefore, the following part will interpret the epigenetic approaches to the mechanisms of body weight regulations as the current or preliminary final point of a research history and will classify it as the latest phase in the genealogy of the genetic and genomic research in this field. In a next step, the focus will be briefly on the possible consequences of the current state of knowledge for public health, more precisely, for the prevention of obesity and the co-morbidities associated with it. According to the stance taken here, the focus will be on an ethical perspective. Prior to that, some conceptual preliminary considerations have to be outlined.

2   Conceptual Preliminary Considerations

The view that the intake of certain food can have positive or negative effects not only for the person eating but also for their descendants is at least in the Judeo-Christian cultural sphere quite common and established within the cultural conscience by famous examples. The most prominent example might be when Adam and Eve were eating the fruit of the Tree of Knowledge in the Garden of Eden (the indefinite fruit from the Bible's text was only turned into an apple by iconography), which is linked to the human tendency towards evil throughout all generations following Adam and Eve by religious or respectively religio-cultural interpretation (cf. Matern in this volume). The most memorable form of this metaphorical transgenerational link can be found in a saying common within the Judean population, which is delivered by the prophet Ezekiel: "The fathers eat the sour grapes, but the children's teeth are set on edge" (Ez 18.2). This formula is especially suitable for the field of epigenetics because while pointing at a transgenerational effect, the question of responsibility is raised at the same time—this is

---

[2]According to the definition of the World Health Organization (WHO 2013), these are diseases which are chronic and non-communicable (e.g. cancer, diabetes, cardiovascular diseases and chronic respiratory diseases).

a problem which, regarding epigenetic lines of tradition, has been the focus of an ethical perspective from the beginning.

Maybe especially in the light of this metaphorically established and recorded link of nutrition and "hereditary" effects, it is hardly surprising that also the scientific foundation of the concept of "epigenetics" is quite substantially based on the field of metabolism and the health disorders also conditioned by nutrition. Epigenetic mechanisms are of particular interest due to their heritability—understood in a certain sense, to put it cautiously. Results such as were obtained from the Överkalix studies or the studies of the F2-generation of the so-called Hunger Winter of the Netherlands primarily draw attention to themselves because apparently it can be shown that the epigenetic changes induced by the nutritional situation of the F0-generation prove to be stable even when the organisms affected are no longer exposed to this nutritional situation (cf. Bode in this volume). It has to be initially clarified, however, in how far such undoubtedly transgenerational effects can actually be traced back to hereditary processes which presume a certain stability of the epigenetic changes also over several generation successions. This has to be done especially because it is also part of the characteristics of epigenetic processes to be reversible and variable (Youngson and Morris 2013).

It is first of all clear that considering the F1-generation, such an interpretation is out of the question since the embryos or respectively the fetuses were directly exposed to the nutritional situation of the F0-generation and thus did not inherit the respective epigenetic profiles in the proper sense. It is not least for this reason that generally the demonstration of the heritability of epigenetic modifications works by connecting the F0-generation and the F2-generation. Nevertheless, even if epigenetic alterations whose causes can be reasonably assumed to lie in the grandparents' generation can be found in the grandchildren's generation, this does not at all prove the heritability of these characteristics in the strict sense. At least it can be argued that not only the F1-generation but also the F2-generation were indirectly exposed to the nutritional situation of the F0-generation in so far as the epigenetic modifications of the F1-generation also affect those cells of which the germ line cells are formed from which the F2-generation will develop. The epigenetic profiles identified in the grandchildren's generation could thus be attributed to an at least indirect exposition which has led to an alteration which is stable for more than one generation, but which can also be regarded as the last effect of the exposition of the F1-generation.

A more detailed discussion of these and similar lines of argumentation can be omitted at this point, since it is primarily important to show that the concept of epigenetics is in need of further plausibility checks to actually be able to be regarded as valid in all its particulars. According to the argumentation demonstrated, the proof of heredity would *stricte dictu* presume that the observed transgenerational effects cannot be ascribed to their own (indirect) exposition of any kind whatsoever, and the epigenetic effects of the nutritional situation of the F0-generation can thus be proved as stable up to the F3-generation. Although indications can be found in respective studies, they are not very many and were found by means of animal model systems so that those findings cannot be simply transferred to human beings. It can be regarded as a hypothesis to be clarified, in how far epigenetic profiles can be understood as stable biological characteristics which are traceable for over two generations and in that sense hereditary and in how far they can thus be explained. The scientific indications are currently at any rate sufficiently numerous in order to base the following considerations on the central assumption of epigenetics—i.e. a heritability, which needs to be further clarified and developed in detail, of effects which do not directly show in the DNA sequence but are preserved in another form across the succession of generations. That is, on the one hand, to characterize obesity research as an area in which epigenetic concepts are gaining relevance and, on the other hand, to outline an ethical perspective towards the field of epigenetics.

3 Body Weight Regulation as a Paradigmatic Field of Epigenetic Research

Obesity is one of the most prevalent and at the same time preventable health risks particularly in the western world (Ogden et al. 2014). Its prevalence in Germany, measured with a BMI $\geq$ 30 kg/m^2, is 23.3 % with men and 23.9 % with women, according to latest data available from Germany. With children and adolescents, the prevalence is 15.0 %. These figures affirm the trend observable for some years now that the prevalences in total either only increase a little or stagnate at a high level. However, there are trends behind these epidemiological observations which give cause for concern. Since the obvious immobility of prevalence rates with adults is caused by a contrary development: Whereas the number of adults with normal overweight (i.e. BMI 25.0–29.9 kg/m^2) decreases, the total number of obese adults increases, as well as the number of extremely obese people (i.e. BMI $\geq$ 35.0 kg/m^2). The latter is particularly also true for children and adolescents (Mensink et al. 2013). What makes obesity a pressing health issue of today is the fact that increased body weight presents one—and quite often the most important —risk factor for a whole range of pathologies, with those of metabolism such as type 2 diabetes mellitus and cardiovascular diseases, but also orthopedic complications, dementia illnesses, certain tumor pathologies or sleep apnea.

Although obesity is a common phenomenon and has to be rated as a priority due to the range and seriousness of its associated co-morbidities by public health and health policy, neither sufficiently effective therapies nor effective strategies of prevention have been developed to date despite various efforts. This fact can be conceptually ascribed to, among other things, the circumstance that despite intensive scientific research an adequate modelling of the obesity phenomenon is still lacking. Whereas from a medical perspective overweight is still primarily an individual issue which has to be solved on the level of the individual, the further growing evidence that obesity has to be equally understood as an epiphenomenon of certain, not at all negative, social developments, which again are accumulating and are affecting the forming of the body weight adversely in combination with certain biological dispositions obtained by evolution. The current trend to apply epigenetic models for the research of body weight regulation can be placed in the context of an increasingly precise, but thus also increasingly complex characterization of the obesity phenomenon. The recourse on epigenetics here marks for the time being the most recent one in a number of approaches to decode the influence of genomic mechanisms to metabolism and the development of body weight in detail. This genealogy has to be illustrated to begin with (concerning this, cf. also Ried 2011).

3.1 Stages of Obesity Genomics

The circumstance that significantly increased body weight can be ascribed to the excessive consumption of nutrients—coupled with a low degree of physical activity—is not only plausible, but also a prevailing assumption which can be initially presumed at least in the sense of common sense. However, there are more than occasionally frequently huge similarities concerning the development of body weight among biological relatives which cannot be simply explained by pointing at an equal behavior in eating and activity. These accumulations of obesity within families could not be explained with the models in which behavioral factors dominated and they drew the attention to possible biological, more precisely, genetic components. At the beginning of the 1990s, formal genetic studies, i.e. twin studies, adoption studies and family studies, were conducted against this background. They especially focused on studying the effects of directed overeating and the development of body weight of twins growing up apart. These studies led to a broad range of estimations of heredity, which ranged from 60 to 90 % in twin studies and from 5 to 70 % in adoption studies and family studies.

Although a more or less important hereditary component of body weight regulation had been established by this, one fundamental problem remained. The pandemic prevalence of obesity across whole populations is such a new phenomenon that it cannot be ascribed to extant mutations within the genome prevalent throughout populations. Therefore, the genomic perspective was flanked and supported by the hypothesis of a *thrifty genotype*. According to this thesis, particularly those genomic constellations have prevailed which can quickly store energy and only emit it thriftily. These genomic profiles which increase the evolutionary biological *fitness* under circumstances of recurring periods of hunger are now, however, facing an environment in which food is not scarce but—at least in the western world—available at any time and in great quantities. Furthermore, huge amounts of calories accumulate in meals due to the industrial processing of food in not at all always advantageous nutrient combinations. These calories are not used (or consumed) by the organism and thus accumulated in its fat depots. The question how this hypothesis can be made plausible on a molecular genetic level and to which factors the variance identified in the formal genetic studies can be ascribed motivated the further development of obesity genetics. Here, initially extremely seldom monogenic forms of obesity were identified which are connected to genetically conditioned disruptions, especially in the leptin system, more precisely, with functionally relevant mutations in the leptin gene or the leptin receptor gene. So far, only very few cases of this monogenic form could be recorded all over the world. In these cases there are also good therapeutic options by administering recombinant leptin.

Main gene effects, of which the previously most relevant one is based on the fundamentals of the melanocortin system, are more frequent than those rare monogenic phenomena. The so-called MC4R receptor is primarily expressed in the hypothalamus and the intestines, and it is functionally assigned to governing the feeling of hunger and satiety. Between 2 and 4 % of the extremely obese people show functionally relevant mutations in the MC4R gene, with male carriers being on average 15–20 kg and female carriers being on average 30 kg more heavy than the reference group. Even more common, but significantly less effective, are different variants in the so-called *Fat and Obesity-associated Gene,* in short: FTO, which is associated with the regulations of the feeling of appetite and satiety. About 16 % of the Caucasian population is homozygous for the risk alleles. The carriers are generally 3–4 kg heavier than the age-adjusted reference group. Their obesity risk is 1.67-times as high. However, there is a phenotypically uncertain tendency that carriers develop a preference for food with a high energy density: They consume more saturated fats by comparison. Especially children show, apart from the stronger inclination to food containing fat, increased *loss of control (LOC)-eating*. Although the relation between FTO variants and macronutrients cannot be called scientifically solid, studies confirm the association to a higher BMI and waist to hip ratio (WHR).

All in all there is a tendency towards those variants with a greater prevalence, and at the same time effect sizes are becoming less and less, but they can accumulate and thus become phenotypically visibly effective. The interindividual variance, however, remains undetermined. The hypotheses from the field of epigenetics take over at this point, with particularly the influence of nutrients for gene regulation, and here especially the interactions of the known susceptibility genes for obesity mainly with processed carbohydrates and saturated fatty acids, being at the center of attention (Dijk et al. 2014; Zerres & Eggermann 2014). Apart from the upward regulation of obesity-associated genes, mechanisms of the downward regulation are also at the focus of interest and thus are the potentially protective effects of certain nutritional profiles (Murray et al. 2014; Tokunaga et al. 2013; Youngson and Morris 2013). Since studies using animal models, such as the well-known agouti mice, have not only shown that epigenetic mechanisms affect the body weight regulation but that those mechanisms can also be levelled out by certain nutritional profiles, epigenetic approaches have evoked growing interest within obesity research. Studies concerning the so-called Hunger Winter of the Netherlands have also indicated that similar mechanisms could be potentially observed with human beings. At any rate, the relevant studies revealed that the generation of grandchildren of those pregnant women affected by the extreme food rationing in the towns and cities of the Netherlands in the winter of 1944/45 are not only smaller and lighter by comparison. The prevalence of metabolic diseases is also raised in the respective genealogical lines and furthermore a lesser degree of DNA methylation is detectable. The following part will briefly outline which implications can result from the growing amount of epigenetic information, for example, for the prevention of obesity.

3.2   Relevance for Prevention

The insights into the effects which certain nutritional profiles have for the epigenetic mechanisms of the regulation of the body weight can become relevant in the context of obesity prevention and associated co-morbidities (Katzmarzyk et al. 2014). Such findings are interesting, for example, if in facilities and institutions in which regularly a larger number of people are catered—e.g. canteens, school cafeterias, refectories etc.—health-promoting nutrient profiles should be offered in the context of an appropriate strategy. On the individual level, findings from epigenetic research are potentially also applicable to nutrition counseling. Furthermore, the response to lifestyle intervention to control body weight can be modulated by epigenetic characteristics. Thus, it is known, for example, that MC4R variants influence the success of *weight loss maintenance* over time. Of course, it needs to be clarified whether epigenetic influences can be noticed here.

Epigenetic profiles could furthermore help to define and identify biomarkers, which (to a certain extent) allow predicting the success of therapies and preventive measures, not least by enabling to differentiate subgroups within a risk collective, for which different measures could be useful. Ultimately, this line runs towards a stratified or respectively personalized prevention which in combination with a *life course approach* particularly addresses those sensitive periods that are especially relevant for regulating body weight. One should here think about the prenatal phase, for example, since the prenatal and perinatal programming have obviously a significant effect on the development of the body weight. However, one has to admit that even without in-depth epigenetic findings, effective prenatal and perinatal obesity prevention is possible, since it is sufficiently known and documented that observing the body weight development of the mother-to-be and her diet has positive effects on the health condition of the child. Here, findings of epigenetic research throw light on previously unknown mechanisms of action and contribute to specify them. It remains undecided for the moment in how far and to what extent this is suitable for increasing the efficiency of the prenatal and perinatal prevention. Furthermore, it must not be ignored that epigenetic knowledge is partly very complex and cannot simply be conveyed into the experience realm of the addressee. For doing so, communicative instruments and strategies are necessary, which again cost degrees of efficiency. Finally, it has to be considered that focusing on nutrient profiles and nourishment profiles can also lead to unilaterally concentrating on the aspect of nutrition ("nutritionism") and thus to drawing the attention in a problematical direction.

All in all, the possible consequences of the epigenetic research for medicine and especially for public health cannot yet be assessed realistically. Besides the questions for possible applications of epigenetics within health care and health protection, there are a number of ethical questions linked to this emerging bioscience. Some highlights are to be discussed in the following part.

4   Ethical Perspectives

One of the most obvious and conceptually most central characteristics of epigenetics is its interdisciplinary focus, in which not only different scientific disciplines
are interlinked, but diverse correlations between scientific disciplines on the one
hand and humane disciplines, cultural science and social science on the other hand
become apparent. This *interdisciplinary* cooperation offers the insight that and the
clarification of how ecological and behavioral factors do not only gain influence (in
the broadest sense) on the level of the physiology of the organism, but furthermore
also on the level of genomics and how they seemingly also keep it over the course
of generations. Thus, it provides the option for an *interdisciplinary* cooperation
affecting both specific and fundamentally conceptual questions. This "bridging
function" is—provided that the mentioned disciplines actually use it as a connecting line—one of the most essential and currently most tangible effects of
epigenetics. This function is also reflected in the unity comprising scientific and
humanistic disciplines. This unity attributes the construction of such an epistemic
as well as epistemological bridge as a task to epigenetics. The editorial title of the
first issue of the relevant academic magazine *Epigenomics* is not chosen by chance
as "Epigenetics: where environment, society and genetics meet" (Majnik and Lane
2014). This concisely sums up the epigenetic program. In the course of the debate
and developments it will have to be observed in how far this encourages a
"molecularisation of biography and milieu" (Niewöhner 2011), which is itself
again and again critically argued especially by social scientists. It will also have to
be observed whether particularly by focusing on and stressing the interconnection
of biological and social factors, greater sensitivity for the complexity of such
interactions is promoted and thus helps reducing one-sidedness.

  Above all, the remaining task will be to discuss the consequences which emerge
if the scientific bridge expected by all is built by epigenetics. Whereas the
scientific-medically oriented discussion regarding the applications and practical
implications of epigenetic research primarily stresses the options for individual-
izing preventive or therapeutic measures, the still rather narrow normatively oriented debate on epigenetics mainly calls socio-ethical perspectives, especially
questions on justice and equality. Thus, on the level of ethical theory the postulate
can be added to suspend the individually centered focus of biomedical ethics
especially coined in North America and Europe in favor of a communitarian focus
(Dupras et al. 2012). At this point at the latest, it becomes obvious that these two
tendencies are not easily made congruent. They form a tense structure which will
accompany and fuel the further discussions not only on the theoretical level.

Specific ethical challenges could emerge from the further consequences of epigenetic research when and in so far as significant biomarkers are identified. Those biomarkers can function as diagnostic instruments for health relevant biological characteristics evoked by outside influences: "there is a *prima facie* duty of justice to intervene [...] when epigenetic information is a reliable biosensor (for) a process of life-time accumulation of disadvantages ending up in disease" (Loi et al. 2013, p. 149). This duty on the one side could correspond to a claim right of those affected on the other side, namely to obtain benefits from the general public as compensation and improvement for their state of health: "Epigenetic screening might diagnose early-life and inherited insults due to factors beyond personal control, thus giving people reasons to demand the provisions of health services as a matter of equality of opportunity" (ibid., p. 151).

Such a positioning, however, gives rise to a number of ethical questions. Initially, it would have to be clarified in how far the availability of biomarkers which have been won on the basis of epigenetic findings can affect the justification of duties and rights. Even without biomedical instruments for diagnostics, there is already sufficient evidence to deduce estimates of probability regarding future diseases or respectively risks from the accumulation of certain social and medical factors. It is not least the body weight which is used as a standard aspect, besides other relevant components. And obesity itself is a clear example for the fact that and how convening different non-genomic factors—nutritional profiles and activity profiles, socio-economic circumstances, residential environment—allow fairly stable statements on probability regarding the risks for the development of increased body weight. It may seem to be a plausible hypothesis if and in how far epigenetic information leads to more valid assessments at this point, but it first and foremost needs to be verified and cannot be simply assumed.

Another aspect which is probably important for the ethical perspectivation of epigenetics and which has also been accompanying the development of obesity genetics from a very early point is associated with the conception of the relationship between personal responsibility and the presupposed significance of genetic data. Biological, inherited factors are paradigmatic for relevant influences that are outside the personal control. Thus, the affected individuals cannot be taken responsible for them. Therefore, genetic findings seem to be appropriate to subvert prevalent stigmatizations and the ascription of responsibilities and to relieve those affected (not only) in a psycho-social manner. In the face of prevailing and socially stable stigmatizations of obesity, an idea is being linked to the emerging genetic or genomic research of body weight regulation. This idea aims at achieving destigmatizational effects, besides the precise clarification of biological mechanisms (Friedman 2004). The insights gained in the context of obesity research point to a general problem area which has to be dealt within the framework of ethical per-spectivation of epigenetics: The question whether genetic—and respectively also epigenetic—information can fulfill these expectations in a reasonable way, cannot be answered without further considerations. For one thing, genetic findings can also be stigmatized so that it has to be initially clarified how different lines of stigmatization are colliding and which effects are achieved by that. For another thing, it remains to be examined whether the hoped-for effects of relieving responsibilities do not presuppose or support a certain, factually not correct and possibly problematic understanding of the genetic findings.

Finally, the focus will be put on the possible function of epigenetic information as *biosensor*: In the context of considerations on justice within the health care system, epigenetics could contribute to more precisely defining the *worst-off* within the group of *worse-off*, which are suffering especially from the health risks and are affected by inequalities, by means of a diagnostics based on biomarkers. Apart from the already illustrated issue of the reliability of such diagnostics, the question arises—especially concerning the aspect of justice—whether it can actually be justified to focus on a certain subgroup in view of the possibility to generate use for a greater specified group of those affected by health risks, even if not in the same way, by social and political measures that are more universally designed. Fur-thermore, it has to remain completely open which specific interventions differing from those generally available are at all qualified if a more accurate diagnostics succeeded with the help of epigenetic knowledge. As long as no differentiated options are available, the usefulness of a division into subgroups concerning epigenetic differences can and must (for now) remain questionable.

5   Conclusion

Substantial potential is ascribed to epigenetics to come closer to a solution for previously pending questions, not only but especially concerning the emergence and transmission of diseases. Even though currently only relatively few specific knowledge could be gained and hardly any applications of epigenetic research could be developed, the questions, which have become the center of attention due to the emerging epigenetics, provide possibilities to more intensified transdisci-plinary cooperation—in the field of body weight regulation and also in other scientific fields: "the field of epigenetics will continue to make the crucial links between environmental and nutritional risk factors and human disease" (Rozek et al. 2014, p. 116). Within the ethical questions that arise alongside the devel-opment of epigenetics, those which need more attention are the decidedly socio-ethical questions in order to anticipate and discuss the social implications of possible more specific consequences already in advance.

References

Bauer, U., Briss, P. A., Goodman, R. A., & Bowman, B. A. (2014). Prevention of chronic disease in the 21st century: Elimination of the leading preventable cases of premature death and disability in the USA. *Lancet, 384*, 45–52.

Berdasco, M., & Esteller, M. (2013). Genetic syndromes caused by mutations in epigenetic genes. *Human Genetics, 132*, 359–383.

Blumenberg, H. (2006). *Beschreibung des Menschen*. Suhrkamp: Frankfurt a.M.

Conn, C. A., Vaughan, R. A., & Garver, W. S. (2013). Nutritional genetics and energy metabolism in human obesity. *Current Nutrition Reports, 2*, 142–150.

Dupras, C., Ravitsky, V., & William-Jones, B. (2012). Epigenetics and the environment in bioethics. *Bioethics, 28*, 327–334.

Friedman, J. M. (2004). Modern science versus the stigma of obesity. *Nature Medicine, 10*, 563–569.

Gluckman, P. D., & Hanson, M. A. (2008). Developmental and epigenetic pathways to obesity: an evolutionary-developmental perspective. *International Journal of Obesity, 32*, 62–71.

Haig, D. (2007). Weisman rules! Ok? Epigenetics and the Lamarckian temptation. *Biology and Philosophy, 22*, 415–428.

Katzmarzyk, P. T., Barlow, S., Bouchard, C., Catalano, P. M., Hsia, D. S., et al. (2014). An evolving scientific basis for the prevention of pediatric obesity. *International Journal of Obesity, 38*(7), 887–905.

Lavebratt, C., Almgren, M., & Ekström, T. J. (2011). Epigenetic regulation in obesity. *International Journal of Obesity, 35*, 757–765.

Loi, M., Del Savio, L., & Stupka, E. (2013). Social epigenetics and equality of opportunity. *Public Health Ethics, 6*, 142–153.

Majnik, A. V., & Lane, R. H. (2014). Epigenetics: Where environment, society and genetics meet. *Epigenomics, 6*, 1–4.

Mensink, G. B. M., Schienkiewitz, A., Haftenberger, M., Lampert, T., Ziese, T., & Scheidt-Nave, C. (2013). Übergewicht und Adipositas in Deutschland. Ergebnisse der Studie zur Gesundheit Erwachsener in Deutschland (DEGS1). *Bundesgesundheitsblatt, 56*, 786–794.

Murray, R., Burdge, G. C., Godfrey, K. M., & Lillycrop, K. A. (2014). Nutrition and epigenetics in human health. *Medical Epigenetics, 2*, 20–27.

Nicolosi, G., & Ruivenkamp, G. (2012). The epigenetic turn. Some notes about the epistemological change of perspective in biosciences. *Medicine, Health Care and Philosophy, 15*, 309–319.

Niewöhner, J. (2011). Epigenetics: Embedded bodies and the molecularisation of biography and milieu. *BioSocieties, 6*, 279–298.

Ogden, C. L., Carroll, M. D., Kit, B. K., & Flegal, K. (2014). Prevalence of childhood and adult obesity in the United States, 2011–2012. *Journal of the American Medical Association, 311*, 806–814.

Ried, J. (2011). Adipositasprävention als normatives Konfliktfeld: Ein sozialethischer Grundriss. In W. Voit, P. Dabrock, J. Ried & J. Uddin (Eds.), *Informierte Selbstbestimmung als Ziel staatlicher Adipositasprävention.* 9–35. Baden-Baden: Nomos.

Rozek, L. S., Dolinoy, D. C., Sartor, M. A., & Omenn, G. S. (2014). Epigenetics: Relevance and implications for public health. *Annual Review of Public Health, 35*, 105–122.

Tokunaga, A., Takahashi, T., Singh, R. B., De Meester, F., & Wilson, D. W. (2013). Nutrition and epigenetics. *Medical Epigenetics, 1*, 70–77.

van Dijk, S. J., Molloy, P. L., Varinli, H., Morrison, J. L., & Muhlhausler, B. S. (2014). Epigenetics and human obesity. *International Journal of Obesity, 34*, 1–13.

van Vliet-Ostaptchouk, J. V., Snieder, H., & Lagou, V. (2012). Gene-lifestyle interactions in obesity. *Current Nutrition Reports, 1*, 184–196.

WHO: World Health Organization. (2013). *Global action plan for the prevention and control of NCDs 2013–2020.* Genf: WHO.

Youngson, N. A., & Morris, M. J. (2013). What obesity research tells us about epigenetic mechanisms. *Philosophical Transactions of the Royal Society B, 368*, 1–12.

Zerres, K., & Eggermann, T. (2014). Genetik und Epigenetik. Erklärungsansätze für (geschlechtsspezifische) Mechanismen der Krankheitsentstehung. *Bundesgesundheitsblatt, 47*, 1047–1053.

Author Biography

Jens Ried is Associate Professor of Ethics at the Faculty of Engineering, Director at the Center for Management, Technology and Society and associate member of the Department of Theology at Friedrich-Alexander-University Erlangen-Nuremberg. He is author and co-author of peer-reviewed articles and papers covering several topics within the fields of ethics of emerging biotechnologies—Dynamics of Hybridisation within the Understanding of and the Approach to 'Life'. Bio-objects and their challenges for the relationship of cultural patterns of orientation and emerging biotechnologies, 2015—and public health—Re-entering obesity prevention. A qualitative-empirical inquiry into the subjective etiology of extreme obese adolescents, 2014. His main research areas are public health ethics, obesity prevention, concepts of health and disease, social ethics and ethics of technology.

Contact: Center for Management, Technology and Society, Nuremberg Campus of Technology, Fürther Str. 246c, D-90429 Nürnberg.

Epigenetics and Original Sin. Theological-Ethical Reflections on Heredity and Responsibility

Harald Matern

Abstract

A theological-ethical approach at the topic of epigenetics may happen in two ways: on the one hand in a cultural-hermeneutical-heuristic way, on the other hand in a constructive way. The here presented contribution unites both perspectives and brings them together in the form of a draft which is obliged to the ethics of responsibility. For this purpose, the first question is which aspects of the topic are referred to in the Christian tradition. Here, the focus is on the one hand on public-medial reception, on the other hand on the structural closeness of some aspects of epigenetics to the Christian doctrine of original sin which is an essential element of theological anthropology. The question about the ethical relevance of this element of the dogma leads to the concept of responsibility. Discussing epigenetics allows for formulating this concept in more detail; at the same time such an approach contributes to demonstrating that the de-moralisation of the public discourse is an essential issue precisely of any theological-ethical position.

H. Matern (✉)
Systematische Theologie/Ethik, Theologische Fakultät, Universität Basel,
Heuberg 12, CH-4051 Basel, Switzerland
e-mail: harald.matern@unibas.ch

© Springer Fachmedien Wiesbaden GmbH 2017
R. Heil et al. (eds.), *Epigenetics*, Technikzukünfte, Wissenschaft und
Gesellschaft / Futures of Technology, Science and Society,
DOI 10.1007/978-3-658-14460-9_14

1   Introduction and Question

1.1   Epigenetics as an Explanatory Model

Current research on epigenetics tries to e.g. shed more light on the individual or generational genesis of certain illnesses. In this context, the way in which these studies are presented by popular media is often rather sensational (see Seitz in this volume). For example, concerning a study on possible predispositions for cardiovascular illnesses or also diabetes mellitus *Spiegel online* opened with the headline[1]: "The womb decides who will be ill".[2] Whereas this study (see Gordon et al. 2012) was about the *pre-natal influences* of environmental factors on identical twins, for other studies the *effects of traumatising experiences of adults* are in the fore (see Uddin et al. 2010). "Traumatised as far as to the genes" was the headline of *Frankfurter Rundschau* on a study on epigenetic changes caused by war experiences.[3]

Even *sexual behaviour* has attracted the attention of epigenetically oriented researchers. A study on monogamy among voles shows how significant the first copulation is (see Wang et al. 2013): the latter, it says, results in epigenetic changes triggering off lifelong monogamous behaviour. Also in this case the presentation by the media makes things look much more unambiguous than actually suggested by the study—"The Chemistry of Monogamy" was the headline of *Süddeutsche Zeitung.*[4] Quite a different approach is meant to give an "Explanation for the Mystery of Homosexuality"—as *Welt* has it.[5] I believe this latter point to be particularly interesting.

For, not only the question about the pros and cons of the legalisation and finally legal equality of homosexual ways of life is an issue which has for some decades been particularly intensively discussed by the (political) public of "western" societies. Rather, together with these questions at the same time an entire

---

[1]All German newspaper headings are translated by the author.

[2]Der Spiegel 17 July 2012: http://www.spiegel.de/wissenschaft/medizin/epigenetik-entscheidet-bereits-im-mutterleib-ueber-krankheitsrisiko-a-844101.html. Last access: 8 May 2016.

[3]FR 9 August 2011: http://www.fr-online.de/wissenschaft/epigenetik-traumatisiert-bis-in-die-gene,1472788,8788360.html. Last access: 8 May 2016.

[4]SZ 4 June 2013: http://www.sueddeutsche.de/wissen/epigenetik-die-chemie-der-monogamie-1.1688030. Last access: 8 May 2016.

[5]Welt 11 December 2012: http://www.welt.de/wissenschaft/article111946147/Forscher-erklaeren-Mysterium-der-Homosexualitaet.html. Last access: 8 May 2016.

anthropology, including a certain idea of human "nature", is at stake which not only attributes the sexual preferences and unambiguous sexual identity of humans to each other but declares this relation to be "natural", the "natural" at the same time being that what is "healthy"—and that what is capable of creating "life". This public debate on dealing with homosexuality was accompanied not only by scientific reflections which, in the form of gender studies, were soon established at universities (on the history of the subject see Frey Steffen 2006; Becker and Kortendiek 2010; Löw and Mathes 2005). Rather, also the natural sciences had been dealing for quite some time with the question of the ("natural") "cause" of homosexuality (for an overview see Horton 1995). This research was driven by genetics. For example, as early as 20 years ago Dean Hamer suggested assuming the existence of a "gay gene".[6] Obviously the public acceptance of homosexual ways of life does not require the destruction but the transformation of the crucial concept of traditional, sexually bipolar and, when it comes to preferences, clearly determined "nature". This has recently been shown by debates in the British parliament. There was no other way to transform into political language and enforce the legalisation of homosexual partnerships than by referring to homosexuality as also being "natural".[7]

Now epigenetic research promises new possibilities for an explanation, like most recently a study by Rice et al. (2012) which met much public response. For the (medial) public reception of such research often the term "explanation" plays a crucial role, followed by "cause". *Scientist* asks: "Can Epigenetics explain Homosexuality?"[8] In *Focus* we could read: "Gene Regulation Causes Homosexuality".[9] The *Standard* headlined: "Possible Explanation for Homosexuality

---

[6]Hamer's studies are based on the research by LeVay (1991), published for the first time in Hamer et al. (1993) (see also King and McDonald 1992). Hamer's study was soon criticized (see e.g. Fausto-Sterling and Balaban 1993).

[7]See Spiegel 17 July 2013: http://www.spiegel.de/politik/ausland/britisches-parlament-erlaubt-homo-ehe-in-england-und-wales-a-911547.html (last access 8 May 2016), Deutsch-landradio 30 July 2012: http://www.dradio.de/dkultur/sendungen/weltzeit/1825907/ (last access 30 July 2014) and Welt 21 March 2012: http://www.welt.de/politik/ausland/article13935142/Selbst-Konservative-kaempfen-fuer-die-Homo-Ehe.html (last access 30 July 2014).

[8]Scientist 1 January 2013: http://www.the-scientist.com/?articles.view/articleNo/33773/title/Can-Epigenetics-Explain-Homosexuality-/. Last access: 8 May 2016.

[9]Focus 20 December 2012: http://www.focus.de/gesundheit/ratgeber/sexualitaet/erotik/tid-28715/forscher-sind-sich-sicher-genregulation-verursacht-homosexualitaet_aid_885680.html. Last access: 8 May 2016.

Found".[10] *Welt* knew: "Researchers Explain the Mystery of Homosexuality". In *Zeit* we could read: "Mum's Poofter, Dad's Lesbian. How Do Humans become Homosexual? A Theory Is Supposed to Explain the Riddle of Same-Sex Love".[11] Given this public reception, there is definitely the question of what this oft-quoted study really says—and what not.

The study does not say that homosexuality is *hereditary*. However, it makes an attempt at explaining the sexual disposition of humans—if "explaining" is really to be understood as a methodical ideal in opposition to "understanding", consisting of causally tracing back a phenomenon to material causes.[12] In this context, the here referred to study implicitly accepts a debated precondition, that is that sexual preference is innate. And also a second one, this time an evolution-biological one, saying that homosexuality is an anomaly in need of explanation, as according to the (elusive[13]) hetero-normative idea it does not immediately serve reproduction.

In short, the results of Rice et al. may be summarized as follows: There are indicators of a considerable degree of the *heredity* of homosexuality which is widespread (not only) among humans. Now Rice and colleagues turn away from the search for a genetic marker which might give evidence to heredity and turn towards an epigenetically oriented way of asking. Their focus is on the foetus's androgene resistance or sensitivity. Usually female foetuses (with a double X chromosome) are androgene (male sexual hormones) resistant, which prevents the development of a male phenotype, whereas (male) XY foetuses react more sensitively. Rice et al. localise the catalyst of androgene resistance or sensitivity with individually different epi-markers causing the development of gender-specific features. These develop during the ontogenesis and are usually dealt with by re-programing processes, so that they are not inherited. In case of homosexuality, the study says, not all of these markers are extinct during the development of the germ cells, thus leading with some foetuses to the development of germ cell

---

[10]Standard 12 December 2012: http://derstandard.at/1353208906647/Moegliche-Erklaerung-fuer-Homosexualitaet-gefunden. Last access: 8 May 2016.

[11]Zeit 14 March 2013: http://www.zeit.de/2013/11/Homosexualitaet. Last access: 8 May 2016.

[12]Wilhelm Dilthey presented this distinction, going back to Kant, as a methodical paradigm to give reason to the autonomy of Humanities from the natural sciences (see Dilthey 1883, p. 81ff.).

[13]The linear connection between homosexuality and reproductive capacity is not necessary by itself. Nevertheless, meanwhile there are also evolution-theory-based, non-monocausal assumptions criticizing this presumption in detail. See Conversation June 2nd 2014: http://theconversation.com/born-this-way-an-evolutionary-view-of-gay-genes-26051. Last access: 8 May 2016.

features contradicting the chromosomal "gender". According to Rice and colleagues, if they last, these sexually antagonistic (SA) epi-markers do not influence a sexually antagonistic phenotype but indeed sexual preference. Which is to say that they prevent the development of heterosexual preference exactly when particularly strong SA epi-markers from the heterosexual parent encounter particularly weak epi-markers, which catalyse sexual diphormism, in the foetus of the other gender. In other words: Under certain circumstances daughters may "inherit" their fathers' sexual preferences and vice versa. This is the epigenetic foundation of homosexual preference. If we like to follow Rice et al. (2012), this preference is innate and can thus be "naturally" explained.

1.2   Ethical Relevance of Epigenetics

To provide more detailed evidence of the theological-ethical relevance of this topic, I would like to make one further step, leading from the general to the particular.

Obviously human behaviour—for way of life must always be addressed this way—does not only influence one's own epigenetics. In case of pregnancy it is the mother's way of life which may cause epigenetic changes with the foetus. Also the parents' behaviour towards babies and little children may cause such epigenetic changes. Against this background, certainly classical questions of the ethics of responsibility must be re-formulated or be formulated in more detail (see Schuol as well as Boldt in this volume). These questions would concern individual responsibility for oneself, but in particular the parents' responsibility towards their unborn or little children.

These questions are extended and become increasingly complicated by the fact that scientifically it has not been finally clarified which epigenetic changes are actually hereditary and which not. This way the question of cross-generational responsibility must be reformulated. Is it that I am "guilty" of my grandchildren's illnesses? Particularly in view of diabetes this would have to be discussed. Or, to open up yet another perspective: Will my care make my little child also become a caring parent, as it is suggested by animal experiment (see Gleason and Marler 2013)?

However, also the question about determinism or indeterminism—often discussed in the context of human genetics—must be newly raised (see Schuol in this volume). If a war trauma does change my epigenetic state, am I then able to reverse this change? Can I influence my sexual disposition? Can homosexuals still be treated?

Certainly, the way in which these questions are formulated is rather eye-catching, and the state of research gives cause to some reservations and differentiations. Among the reservations there counts on the one hand that only the tiniest number of human phenomena have been researched much or across the generations. Furthermore, there is the fact that only very few epigenetic changes stay stable over the generations, i.e. that they resist the usual mechanisms of re-programming. The question is, which are they and under which circumstances? Finally, one grave mistake suggested by simplifying presentations must be avoided: Only in very few cases there are mono-linear relations between genotype and phenotype. This holds even more for epigenetics. And just like the latter precisely points out that, due to the complexity of the factors contributing to the phenotypical expression of a feature, any genetic determinism is obviously a wrong assumption, it would be fatal to believe, given a new "magic formula", that now the "cause" of homosexuality, diabetes or post-modern fathers has been identified.

Nevertheless, due to these reservations we cannot deny the ethical questions which, given epigenetic research, must be newly and more differentiatedly raised. For, although we must proceed most carefully to avoid premature or undercomplex conclusions, nevertheless in some cases they allow for at least very exactly naming one factor contributing to a complex causal connection.

2   Theological-Ethical Approaches

A theological-ethical perspective must also take the above described ethical questions into consideration: From the stock of traditions of the Christian culture of symbols and reflections it may pointedly contribute to debating them, by e.g. positioning itself as an ethics of responsibility while universalising the forum of responsibility by taking the "acceptable before God" point of view.

Such a position will do complete justice to its obligation of being communicatively connectable only if it reflects on its own positionality. For this, the historical-analytical view at the traditional stocks of symbols and the culture of their reflectivity may be helpful. In the context of the interdisciplinary debate, such a reflective re-appropriation of one's own normative premises may also serve for underlining the independence of the theological-ethical contribution to the discourse—and for formulating this contribution not only as taking a position but furthermore as a hermeneutical way of shedding light on the topics of the discourse.

In our case such a kind of approach suggests itself for the symbolical-theological concept of "original sin".[14] The Christian doctrine of (original) sin (if stripped of its symbolical and partly moralistic guise) reflects those factors as contributing to the possibility of action. In this context it takes two things into consideration: On the one hand, the irreducibility of individual dispositions to act. On the other hand, the fact that the individual is responsible for them, i.e. the original sin is at the same time individual "guilt". Any Christian ethics, at least any Protestant ethics, must also consider the doctrine of sin as an essential component of Christian anthropology. The reference to the doctrine of original sin suggests a structural connection to those questions as being raised by the research on epigenetics.

In the following I will at first shortly discuss what Christian tradition means by original sin (see Gross 1960–1972, Sect. 2.1). Then I will try, by explaining a few points, to work out the relevance of this theorem or "theologumenon" for theology-based ethics (Sect. 2.2). Finally, from a theological-ethical point of view, I will make an attempt both to take a position (Sect. 2.3) and to formulate some general remarks (Sect. 2.4) which might be of relevance also for non-theologists.

2.1 What Is Original Sin?

The Biblical foundations of Christian tradition assume that the first human couple in the Garden of Eden committed an *original* sin: Eve gave to Adam a fruit from the Tree of Knowledge, and Adam ate it. Previously God had banned them from eating it. The result, part one: Adam and Eve realise their nakedness and feel ashamed. The result, part two: Adam and Eve are expelled from the Garden of Eden and from then on must suffer from the burden of labour and reproduction as well as mortality.[15]

Early Christian tradition commented differently on this "original sin", most of all concerning its material content.[16] In this respect, initially there were two candidates. The first one was gluttony: eating when there is no need to eat, that was Adam's sin. The second one was sexuality. The general idea that the actual offence was disobedience to God pushed through only at a later time.

---

[14]The closeness of both conceptions to each other has also already attracted the attention of others. See Hughes (2014).

[15]See the Biblical narration of Paradise Gen 2, 4b–3, 24 (on this Dohmen 1988).

[16]On the genesis of the doctrine of original sin in the Ancient Church see Beatrice (2001) and Scheffczyk (1981).

Then, in the first two centuries AD there is the idea that Adam's sin was handed
over to Adam's descendants by way of the sexual act, i.e. reproduction—and from
them to their descendants and so on. These are the two main aspects of the concept
of original sin. In German, the difference between the original act and the heredity
of its (dispositional) effects in the following generations is marked by the use of
two different terms: "Ursünde" (meaning the original act) and "Erbsünde"
(meaning the handing over of its effects). According to the idea of "Erbsünde", all
humans are born with a "sinful" disposition. The question of what now this means
may be answered differently: Most theologians interpret this sinful disposition as a
basic flaw, entailing the incapability to do good things and, furthermore, being
expressed by a tendency towards revolting (against God)—both as a tendency
towards the kinds of behaviour of "greediness" and of "gluttony" as well as others
which are traditionally called "deadly sins".

Here I would at first like to support a thesis, to then extend the topic. Both will
serve for understanding the specifically theological-ethical approach at the topic in
more detail.

1. It is no coincidence that (epi-) genetic research deals with diabetes and
   homosexuality. Obviously "gluttony" and "greediness" are
   cultural-anthropologically important ways of behaviour which, at least in our
   culture, are often connected to moral irresponsibility. Still today the food
   chemist Udo Pollmer may metaphorically call appetite "the modern original
   sin".[17] Overindulging parents are guilty of their descendants having this disposition. A homosexual preference must somehow be originally connected to
   "sex". Much scientific effort is made to give evidence to such guilt, as well as to
   liberate the descendants from it. It is part of the elucidating tasks of theological
   ethics to point out to the possible symbolic backgrounds of such perceptions
   and to explain them in a differentiating way. This is a *cultural-hermeneutic*
   contribution to the debate.
2. The complex of symbolic interpretation introduced by the doctrine of original
   sin refers to the anthropological constant of innate and thus "inevitable"
   physical and behavioural dispositions. This is neither meant to say that the
   doctrine of original sin could be "proven" by epigenetic research nor that
   certain inherited dispositions must reasonably be described as "sins". Rather, I
   would like to direct attention to the fact that since the appearance of the
   theorem of the doctrine of original sin in the New Testament (St. Paul) there

---

[17]Zeit 11 July 2013: http://www.zeit.de/lebensart/essen-trinken/2013-06/ernaehrung-diaeten.
Last access: 8 May 2016.

belongs something more. Not only an innate behavioural disposition is described there but also a specific way of dealing with it is suggested. For, the Christian tradition says, the original sin is always also individual guilt. Thus, one must take *responsibility* for it, as if it was the result of one's own behaviour. Parallel to the doctrine of original sin there developed theological ethics providing different models of dealing responsibly with the inherited disposition. In the course of the history of Christian theology one developed both very differentiated ideas of how exactly one might imagine the heredity of behavioural dispositions and of which ways of dealing with it exist. Providing and "translating" these models is the contribution of theological ethics to the debate.

2.2 Original Sin and Theological Anthropology

Theological anthropology, in particular the Protestant one, is basically dominated by anthropological pessimism. The human being is a sinner, from the cradle to the grave, i.e. of his/her own accord he/she is incapable of intending what is good. He/she is also incapable of doing it. This is the dispositional aspect when it comes to individual *behaviour*. At the same time the human being is physically incomplete. It is prone to illness and mortal. Furthermore, both self-preservation and reproduction are laborious. Both require both physical labour and sometimes also exhausting pregnancy and painful birth. This is the dispositional aspect of original sin when it comes to the *physical* disposition. Original sin is general. However, at the same time it is individual. Each and every one is different, but all must bear a "blemish".

Now, the question is where these flaws come from. The theological tradition knows different models of explanation. They are surprisingly differentiated and, particularly in the modern age, additionally characterised by a biological component, even by "systemic" aspects. I would like to give some examples of this:

(1) The human being is born with a psychic (not: physical) flaw because God, after having created physical life, made the individual soul out of nothing, while nevertheless it bears original guilt. At the same time this guilt gives reason to the humans' ability to carry responsibility.[18]

---

[18]This position was supported a. o. by reformed Orthodoxy. See the compilation in Heppe (1861, pp. 231). On how the concept of sin was understood by Protestant theology in the Modern Age see in general Axt-Piscalar (2001).

(2) Each human inherits his/her sinful disposition directly (physically) from his/her parents.[19]

(3) Individual behavioural disposition results from habitualisation processes in the course of individual biography.[20]

(4) Behavioural dispositions reflect different possibilities of accepting the contingency of the individual freedom to act together with its legal necessity. At the same time this contingency includes a previous determinedness of individual freedom.[21]

(5) Behavioural dispositions result from social structures and thus connected (economic) constraints[22]—this way they also refer to ontological "alienation processes".[23]

These theological models of inheritance and heredity show very well the different possibilities of describing the existence of certain dispositions. Taken together and as a cross section, they produce an image which, perhaps precisely because of its complexity, allows for approaching the phenomenon of inherited epigenetic effects.

Individual psychic and physical dispositions are partly biologically inherited. Perhaps not over generations, but at least directly from the biological parents. Their habitualisation or shape requires certain environmental influences or also certain ways of behaviour in the course of the individual biography. Also each individual is responsible for this aspect, at least in so far as it must take the burden. However obviously, due to the complexity of the interplay of psychic and physical predispositions, purposeful self-influencing is often psychically and in many cases also physically impossible. In this context, often the psychic and physical possibilities, in view of the individual biographic development, are inversely correlatively related to each other. Furthermore, those social and economic institutional contexts within which individual biographies develop play a relevant role for this complex image.

Taken together, here results a kaleidoscope of different aspects influencing the heredity and also the persistence of physical and psychic dispositions. In so far as

[19]This position is found among the Arminians (see Seeberg 1969, pp. 662; Hoenderdaal 1979).

[20]This is the position of the Socinians (see Urban 2000).

[21]This position was supported—in different ways respectively—by Immanuel Kant and Johann Gottlieb Fichte (on this: Fischer 1988; Fittbogen 1907; Gestrich 1995).

[22]See Ritschl (2002, §39); on this: Axt-Piscalar (1996, pp. 271).

[23]See Tillich (1958, p. 54); on this Wenz (1996) and Danz (2000, pp. 211).

none of them can be telling on its own and all of them taken together are hardly comprehensible, one may speak of "inevitable" dispositions.

From this theological ethics draw the conclusion that the individual, due to its incapability of overcoming its dispositions by itself, is in need of help. Theologians call this "justification" and "redemption". These concern, on the one hand, the individual's salvation, but on the other hand also the attempt at restoring physical health as well as the biggest possible correlation of intention and action when it comes to the individual lifestyle. Precisely concerning the latter aspects, not only the individual's relation to God is of significance but in particular the ways in which they are mediated. They exist within social institutions: Family, educational system, political community, but also medicine and the sciences. Any theological ethics of responsibility must consider itself against this background.

2.3   Theological Ethics of Responsibility

Roughly, one may say that any theological ethics of responsibility taking a position towards the insights of epigenetic research without ignoring its own stock of symbolic traditions must take the following dimensions of responsibility into consideration:

1. the level of the individual: Each individual must be responsible for his/her own life;
2. the social level: Responsibility does not only cover myself but just the same all other fellow humans. On the one hand in a way that my life will be able to bear social burdens; on the other hand while considering the fact that fellow humans might depend on my support. This holds also cross-generational and is substantiated by the institution of family. Its members take over responsibility for themselves, but also for each other. And this happens cross-generational and asymmetrically, responsibility taking the opposite direction in the course of one generation. Parents bear the responsibility for their children, both in view of their lives being as healthy as possible and of the possibilities of psychic self-regulation, by making emotional care and educational possibilities possible in a broad sense. Vice versa, children will take the responsibility for their ageing parents as well as for their own children.

In this context, precisely the awareness of general, however individually different "sinfulness" allows for recognising and respecting the limits of both one's own and other's responsibility, just like the limits of one's own capability to act. At the level of the individual, the family and the social level in the broader sense, not much is in our hands, both in the psychic and the physical respect. Nevertheless, everything possible is supposed to be done. Of this, what now is of interest for an interdisciplinary debate?

2.4   Epigenetics and Original Sin

Both from a cultural-hermeneutic-heuristic and from a constructive point of view, theological ethics are capable of providing an independent and possibly connectable contribution to the ethical debate on epigenetics. On this, I would like to formulate some theses.

(1)   From a *heuristic* or a *cultural-hermeneutic* point of view, a look at the Christian tradition of the doctrine of original sin may be helpful for sharpening our awareness of the fact that starting points of epigenetic research—I have referred to the examples of adiposity/diabetes—are possibly not always due to scientific interest but also to un- or semi-conscious *moral* judgements already resonating with the way of formulating the question—as well as and particularly when it comes to the way in which the media respond to this research. Often these judgements imply the idea of *total responsibility*— either from an individual or a generational perspective. From this point of view, both would have to be moderated and stripped of their moral implications which may possibly be considered, among others, to be the culturally determining effect of a moralistic narrow-mindedness of the Christian doctrine of original sin itself. In this respect, theology has particular *obligations to explain*.

(2)   From a *constructive* point of view, the explanation offered by theological ethics may precisely consist of pointing out to the above sketched relational complexity of responsibility. It must on the one hand be sharpened, on the other hand it must be moderated concerning its total validity. Given the complexity of their development, individual dispositions are inevitable. Indeed, everything possible should be done to avoid harmful dispositions. However, knowing that this is not completely in our hands means also that we must learn how to live with them, i.e. with the dispositions of our fellow humans, as far as ever possible. In this context, asking for help from others does not only make things easier but is indeed inevitable.

From this we may draw several conclusions: On the one hand, the already existing infrastructure should be extended. Both in view of public *education*, i.e. in our case providing information about heredity, and this means both positive and negative aspects: What shall we do or not do? What are we able to influence and what not? Such an educational offer can only partly be provided by public institutions. Concerning this, in particular families (whatever their constellation may be) must be committed but also supported. Precisely given the fact that obviously neither our parents nor we ourselves nor the society we live in are capable of taking total responsibility for both for ourselves and for others, indeed that they are not supposed to do so and do not need to do so (total responsibility is non-existent among humans, it exists only "in the face of God"), from a theological point of view it is imperative to stay away from *moralising* certain dispositions or ways of behaviour, may they be illnesses such as adiposity/diabetes, may they be certain sexual preferences deviating from an assumed "human nature". Instead, we should rather learn how to live with individual differences there where they do not restrict the possibilities to live of those being in need of protection and help. This *de-moralisation* may contribute to more pragmatically starting the extension of existing, supporting infrastructures, without at the same time overestimating the possibilities they offer.

(3) Epigenetic *research*, in particular on the heredity and persistence or mani-festations of individual physical and behavioural dispositions of humans, must be extended. The more exactly we are able to describe the mechanisms of heredity, the more effectively and responsibly public information, diag-nosis, prevention and, if necessary, also therapy may be provided. At the same time, this approach allows for the de-moralisation of particular dispo-sitions as well as for the precise restriction of the responsibility but also the ability to carry responsibility of individuals and social institutions.

In this context, particularly concerning the (media) interpretation of the respective research results, hermeneutical prudence is imperative, and par-ticular attention should be paid to the actual validity of the respective studies.

(4) Finally, precisely due to the above described results of epigenetic research, once again the significance of the way of life and, in the broader sense, of individual possibilities of education move into the focus. This way there also develops an *extension* of the concept of responsibility beyond the physical frame which, with the above given qualifications, should be depicted par-ticularly also when it comes to parental care in early life stages as well as to educational institutions.

References

Axt-Piscalar, C. (1996). *Ohnmächtige Freiheit. Studien zum Verhältnis von Subjektivität und Sünde bei August Tholuck, Julius Müller, Sören Kierkegaard und Friedrich Schleiermacher.* Tübingen: Mohr.

Axt-Piscalar, C. (2001). Art. Sünde VII. In G. Müller, H. Balz, & G. Krause (Eds.), *Theologische Realenzyklopädie.* Bd. 32 (pp. 400–436). Berlin u.a.: de Gruyter.

Beatrice, P. F. (2001). Art. Sünde V. In G. Müller, H. Balz, & G. Krause (Eds.), *Theologische Realenzyklopädie.* Bd. 32 (pp. 389–395). Berlin u.a.: de Gruyter.

Becker, R., & Kortendiek, B. (Eds.). (2010). *Handbuch Frauen- und Geschlechterforschung: Theorie, Methoden, Empirie.* Wiesbaden: VS.

Danz, C. (2000). *Religion als Freiheitsbewusstsein. Eine Studie zur Theologie als Theorie der Konstitutionsbedingungen endlicher Subjektivität bei Paul Tillich.* Berlin u.a: de Gruyter.

Dilthey, W. (1910). Der Aufbau der geschichtlichen Welt in den Geisteswissenschaften. *Abhandlungen der Preußischen Akademie der Wissenschaften. Philosophisch- Historische Klasse, Jg.* 1910, 1–123.

Dilthey, W. (1914). *Weltanschauung und Analyse des Menschen seit Renaissance und Reformation. Abhandlungen zur Geschichte der Philosophie und Religion (GS II).* Stuttgart: Teubner.

Dohmen, C. (1988). *Schöpfung und Tod. Die Entfaltung theologischer und anthropologischer Konzeptionen in Gen 2/3.* Stuttgart: Kath. Bibelwerk.

Fausto-Sterling, A., & Balaban, E. (1993). Genetics and male sexual orientation. *Science, 261,* 1257.

Fischer, N. (1988). Der formale Grund der bösen Tat. Das Problem der moralischen Zurechnung in der praktischen Philosophie Kants. *ZPhF, 42,* 18–44.

Fittbogen, G. (1907). Kants Lehre vom radikalen Bösen. *KS, 12,* 303–360.

Frey Steffen, T. (2006). *Gender.* Leipzig/Stuttgart: Reclam.

Gestrich, C. (1995). *Die Wiederkehr des Glanzes in die Welt. Die christliche Lehre von der Sünde und ihrer Vergebung in gegenwärtiger Verantwortung.* Tübingen: J. C. B. Mohr (Paul Siebeck).

Gleason, E. D., & Marler, C. A. (2013). Non-genomic transmission of paternal behaviour between fathers and sons in the monogamous and biparental California mouse. *Proceedings of The Royal Society of Biology, 280,* 20130824.

Gordon, L., Joo, J. E., Powel, J. E., Ollikainen, M., Novakovic, B., Li, X., et al. (2012). Neonatal DNA methylation profile in human twins is specified by a complex interplay between intrauterine environmental and genetic factors, subject to tissue-specific influence. *Genome Research, 22*(8), 1395–1406.

Gross, J. (1960–1972). Geschichte des Erbsündendogmas. Ein Beitrag zur Geschichte des Problems vom Ursprung des Übels, 4 Bde., München/Basel: Reinhardt.

Hamer, D. H., Hu, S., Magnuson, V. L., Hu, N., & Pattatucci, A. M. L. (1993). A linkage between DNA markers on the X chromosome and male sexual orientation. *Science, 261,* 321–327.

Heppe, H. (1861). *Die Dogmatik der evangelisch-reformierten Kirche.* Elberfeld: Friderichs.

Hoenderdaal, G. J. (1979). Art. Arminius, Jacobus/Arminianismus. In G. Müller, H. Balz, & G. Krause (Eds.), *Theologische Realenzyklopädie*. Bd 4 (pp. 63–70), Berlin u.a.: de Gruyter.

Horton, R. (1995). Is homosexuality inherited? The New York Review of Books, July 13. http://www.nybooks.com/articles/archives/1995/jul/13/is-homosexuality-inherited/. Last access March 2016.

Hughes, V. (2014). The Sins of the Father. *Nature*, 507, 22–24 [Die deutsche Version erschien online unter dem Titel „Vaters Erbsünde" in *Spektrum*. http://www.spektrum.de/news/vaters-erbsuende/1258600. Last access March 2016].

King, M., & McDonald, E. (1992). Homosexuals who are twins: A study of 46 probands. *The British Journal of Psychiatry, 160*, 407–409.

LeVay, S. (1991). A difference in hypothalamic Structure between heterosexual and homosexual men. *Science, 253*, 1034–1037.

Löw, M., & Mathes, B. (Eds.). (2005). *Schlüsselwerke der Geschlechterforschung*. Wiesbaden: VS.

Rice, W. R., Friberg, U., & Gavrilets, S. (2012). Homosexuality as a consequence of epigenetically canalized sexual development. *The Quarterly Review of Biology, 87*, 343–368.

Ritschl, A. (2002). *Unterricht in der christlichen Religion, Studienausgabe nach der 1. Auflage von 1875 nebst den Abweichungen der 2. und 3. Auflage. Einl. u. hrsg. v. Christine Axt-Piscalar*. Tübingen: Mohr.

Scheffczyk, L. (1981). *Urstand, Fall und Erbsünde. Von der Schrift bis Augustinus (HDG II/3a/1)*. Freiburg u.a.: Herder.

Seeberg, R. (1969). Lehrbuch der Dogmengeschichte. Vierter Band, zweiter Teil: Die Fortbildung der reformatorischen Lehre und die gegenreformatorische Lehre. Vierte Auflage (=photomechanischer Nachdruck der dritten Auflage. Darmstadt: Sonderauflage der Wissenschaftlichen Buchgesellschaft.

Tillich, P. (1958). Systematische Theologie, Bd. II. Stuttgart: Evangelisches Verlagswerk.

Uddin, M., Aiello, A. E., Wildman, D. E., Koenen, K. C., Pawelec, G., de los Santos, R., et al. (2010). Epigenetic and immune function profiles associated with posttraumatic stress disorder. *PNAS, 107/20*, 9470–9475.

Urban, W. (2000). Art. Sozzini/Sozinianer. In G. Müller, H. Balz, & G. Krause (Eds.), *Theologische Realenzyklopädie*. Bd. 31 (pp. 598–604). Berlin u.a.: de Gruyter.

Wang, H., Duclot, F., Liu, Y., Wang, Z., & Kabbaj, M. (2013). Histone deacetylase inhibitors facilitate partner preference formation in female prairie voles. *Nature Neuroscience, 16*(7), 919–924.

Wenz, G. (1996). De causa peccati. Die Lehre vom Urfaktum der Sünde in Paul Tillichs Systematischer Theologie. In Viertel, M. (Eds.): *Gott und das Böse* (pp. 9–33). Hofgeismar: Evangelische Akademie.

Author Biography

Harald Matern Dr. theol. is research fellow and lecturer of Systematic Theology at the Chair of Systematic Theology/Ethics of the University of Basel (Switzerland). His recent works are as follows:
    Personalized Healthcare—Focus on Individuality, in: Dabrock, Peter; Braun, Matthias; Ried, Jens (Hg.): Individualized Medicine between Hype and Hope. Exploring Ethical and Societal Challenges for Healthcare, Erlangen 2012, S. 51–77; Wertgefühle und gelebte Moral. Rudolf Ottos Begründung der Ethik im Anschluss an Kant, in: Lauster, Jörg; Schüz, Peter; Barth, Roderich; Danz, Christian (Hg.): Rudolf Otto. Theologie - Religionsphilosophie - Religionsgeschichte, Marburg 2014, S. 391–402; (with Georg Pfleiderer) (Eds.): Theologie im Umbruch der Moderne. Karl Barths frühe Dialektische Theologie, Zürich 2014.
    His main research interests include the history of modern protestant theology, philosophy of religion, theology of creation and bioethics, and theological eschatology.
    Contact: Systematische Theologie/Ethik; Theologische Fakultät, Universität Basel, Heuberg 12, CH-4051 Basel.